W9-BBD-507

Saint Peter's University Library
Withdrawn

Saint Peter's University Library
Withdrawn

*Coca Exotica*

# *Coca Exotica*

THE ILLUSTRATED STORY OF COCAINE

*Joseph Kennedy*

RUTHERFORD ● MADISON ● TEANECK
FAIRLEIGH DICKINSON UNIVERSITY PRESS

NEW YORK ● LONDON ● TORONTO
CORNWALL BOOKS

© 1985 by Associated University Presses, Inc.

Associated University Presses
440 Forsgate Drive
Cranbury, NJ 08512

Cornwall Books
25 Sicilian Avenue
London WC1A 2QH, England

Cornwall Books
2133 Royal Windsor Drive
Unit 1
Mississauga, Ontario
Canada L5J 1K5

**Library of Congress Cataloging in Publication Data**

Kennedy, Joseph, 1948–
Coca exotica.

Bibliography: p.
Includes index.
1. Cocaine habit—History. 2. Cocaine—History.
I. Title.
HV5810.K46 1985 362.2′93 82-45861
ISBN 0-8386-3103-7 (Fairleigh Dickinson University Press)
ISBN 0-8453-4778-0 (Cornwall Books)

Printed in the United States of America

*To my parents*

# Picture Credits

Sources of maps, photographs, and illustrations are given below.

Sir James Barrie, O. M., 82; Bavaria-Verlag, 114; Mrs. Hortense Koller Becker, 68, 71, 78; Theodore de Bry, 37; *Bulletin on Narcotics,* 26; 31, 115; The Fitz Hugh Ludlow Memorial Library, 15, 36, 46, 49, 51, 56, 58, 63, 64, 81, 82, 83, 84, 85, 86, 87, 108, 112, 113; John Frampton, 43; *Hamptons Magazine,* 89, 92, 93; Dr. Joel Hanna, 129; *Journal of the American Medical Association,* 90, 97; *Jungle Memories* by H. H. Rusby, 1937, used with permission of McGraw-Hill Book Company, 74; Joseph Kennedy, frontispiece; John A. P. Kruse, 14, 114; Rev. John Lamond, D.D., 81; Wing Commander Asher Lee, 113; Loren McIntyre, 21 and color insert (six plates); *Medical Record,* 86; A. A. Moll, 38; *New York Times,* 95, 100; Northwestern University, Archibald Church Medical Library Portrait Collection, 75; Darrell Orwig, 17, 18, 19, 25, 28, 42, 70; Felipe Guamán Poma de Ayala, 33; F. J. Pohl, 32; James Powell, color insert (five plates); Howard Pyle, 45; E. Riou, 22, 30; Trinity College Library, 44; Universal Studios, 118; Arturo Wesley, color insert (three plates).

# Contents

# Acknowledgments

I began this project more than eight years ago at the suggestion of Dr. Mark D. Merlin of the University of Hawaii. From that time until now he has helped and supported me in my efforts to make this book successful. A large debt of thanks is due to Dr. Dennis T. P. Keene, who has served me as friend and patient critic through some confusing and difficult times; Dr. Keene is certainly very familiar with the text of this book by now, for I must have reviewed it with him several hundred times. I owe a special debt of gratitude to all the artists whose illustrations appear in this work. Loren and Sue McIntyre of Arlington, Virginia, contributed the very fine color pictures that have been reproduced here; their sensitive treatment of the subject matter and the requests of a novice author are greatly appreciated. Mr. Daryll Orwig's original art work speaks for itself; throughout the production of these pictures, Mr. Orwig provided a receptive ear as well as a talented imagination and pen. Mrs. Hortense Koller Becker, the daughter of Dr. Carl Koller, has been as cooperative and cordial as anyone I have dealt with in the course of this project and her photos of her father, Freud, and Fleischl are greatly appreciated. The majority of the rare photographs in this text were assembled with the friendly help and permission of Dr. Michael Aldrich and the staff of the Fitz Hugh Ludlow Memorial Library in San Francisco, California. Besides acquiring some fine old photographs in San Francisco, I also received a practical and painful lesson in the politics of drug research. Mr. Bruce Erickson did most of the black and white copy photography work in this book. He also helped salvage some of the photos I took in San Francisco and was of invaluable technical assistance in other matters as well. Thanks are extended to my old friend James Powell who literally crossed the Andes to take photos of the coca *finca* that appear in this book. Mahalo to John Kruse for his coca maps and friendship. The typing involved in this project was done by Jill Kajikawa and Barbara Jagielski, who worked long hours to complete it. I would like to thank Victoria Nelson for her editorial assistance. A special thanks to Mrs. Margaret Smith and the staff of the Hamilton Library at the University of Hawaii; this talented group provided me with invaluable library assistance throughout the project.

The following also deserve mention here and my thanks. They include Kent Davenport, M.D., Karynne L. Chong, Alan Newman, M.D., Dr. P. Bion Griffin, Dr. R. Clark Mallam, Tim V. Dougherty, Richard and Melinda Moulton, Jerry Hopkins, Dr. Joel Hanna, Ken Cassman, Lydia Nakajima, Dr. Marion Mapes, Art and Laurie Sackler, Arturo Wesley, Robert Miller, and Bob Goodman. There is a long list of others.

JOSEPH KENNEDY
HONOLULU

*Coca Exotica*

# 1

# An Ancient Beginning

It has been said that, without coca, there would be no Peru. Although this comment was made by a Spanish conquistador more than 400 years ago, the feeling it represents is still alive today. A coca branch is incorporated into the national emblem of the modern-day Republic of Peru, and the leaf is also displayed on every piece of Peruvian currency. Coca is a physical and spiritual elixir for thousands of Andean Indians, and the cultivation of the leaf in South America has been a significant political and economic factor for the past 5,000 years.

Despite its renown in Peru, the coca leaf has a history of being confused with such unrelated items as cocoa, coconut, and betel. Besides this, there has traditionally been a great deal of confusion surrounding the effects of coca's mildly stimulating psychoactive properties and the history of the leaf's associations with different cultures at different times.

Since the time the cocaine alkaloid was separated from the leaf in 1860, the coca plant has grown to become a substance of great international importance affecting the lives of millions of people. In light of this development there is a need to view coca and cocaine from a complete historical perspective. To achieve this, it is necessary to begin with an archaeological reconstruction of ancient Peru as it existed at the time man first arrived.

The story of man and coca really begins in Peru about 20,000 years ago, when groups of hunting and gathering people first immigrated into the central Andes of South America. During this early collecting period, man must have experimented with literally thousands of plants in his search for food. The nomadic existence that characterizes all hunter and gatherer groups undoubtedly led these small family bands into a variety of econiches throughout the Andes and exposed them to countless new varieties of plants wherever they went. Of all the areas within this part of the world, the eastern, or Amazonian slopes of the Andes, have the greatest diversity of plant life and would, therefore, be the most attractive to fruit and vegetable gatherers. This lush, tropical region is called the *montaña* in Peru, or the *yungas* in Bolivia, and is most likely the home of the coca plant.*

It is reasonable to assume that coca was first sampled by these early hunting and gathering people of Peru when they reached the *montaña,* perhaps as early as 15,000 years ago. It is always difficult to produce "proof" of the associations between wild plant species and man, especially within the natural range of these plants where the earliest experimentations must have taken place. Nevertheless, a plant with the remarkable properties of coca could hardly be overlooked by

*The actual location of coca's wild ancestors is unknown because no one has ever found a truly wild example of the plant. This is a common problem for botanists who study plants that have been under cultivation for long periods of time, e.g., corn, the opium poppy, and, obviously, coca. Recently, however, Dr. Timothy Plowman of the Field Museum in Chicago, Illinois, has made significant strides toward the identification of a truly wild species of coca.

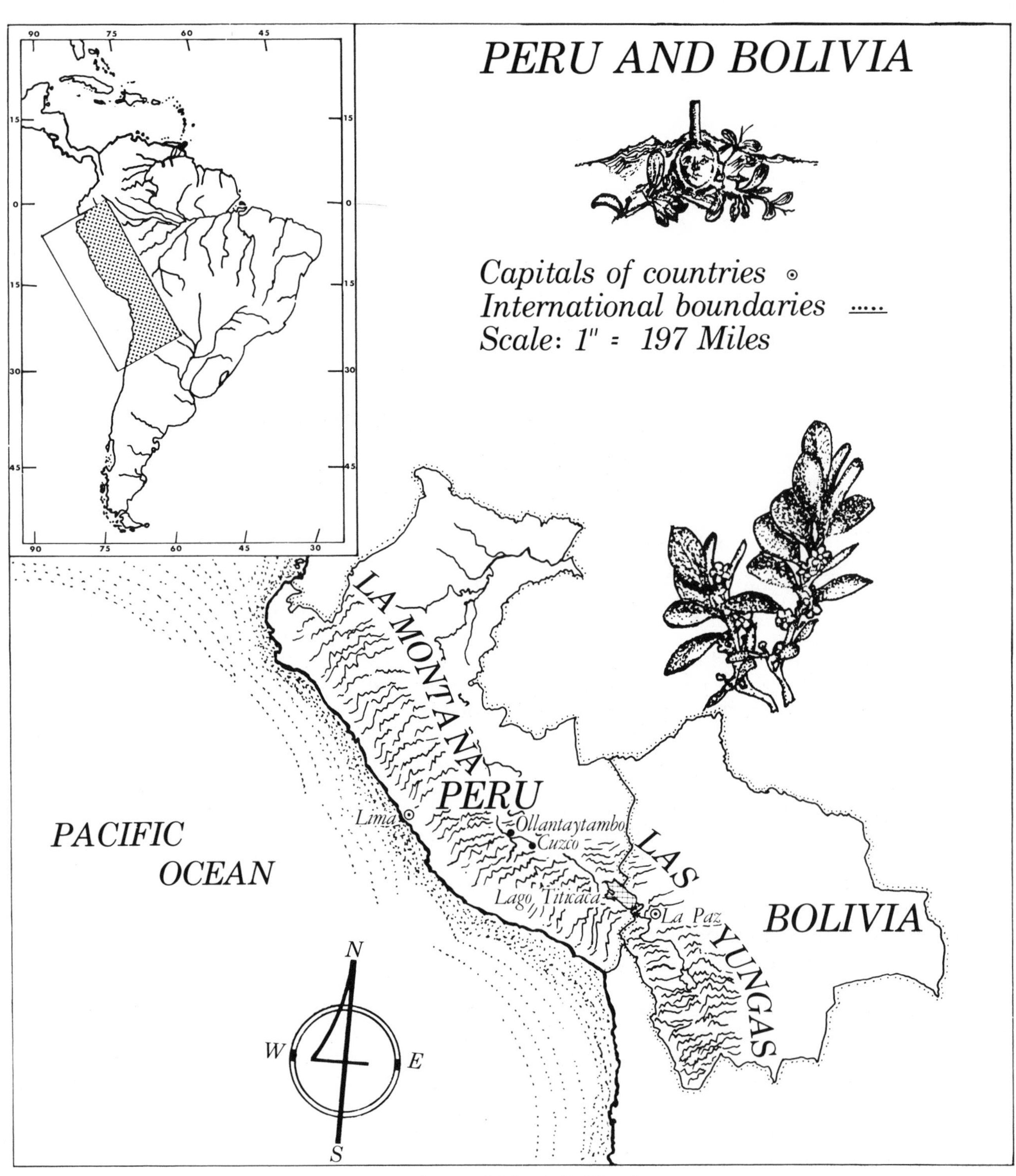


(Courtesy of John A. P. Kruse)

collecting bands, even at such an early date. The most rudimentary testing of the plant—that is, introducing the leaves into the mouth—might be enough to numb the sting of a cut lip or deaden the pain of a common toothache. After being alerted to these obvious benefits, these knowledgeable plant experimenters probably learned that coca could be chewed to combat hunger, cold, and fatigue or infused to remedy stomach disorders. Because such unpleasant physical states were inescapable facts of life for these Andean hunters and gatherers, we may speculate that coca may have been a valuable and well-known member of their ever-changing plant world.

Early man's relationship with coca changed around 2000 B.C., when he learned to cultivate plants and domesticate animals. A welcomed by-product of these developments was surplus, something unknown to the hunter-gatherer. Surplus was stored goods and stored goods meant that more time could be spent on pursuits other than food production. By 2500 B.C., man began to convert this time into a variety of creative cultural pursuits, resulting in a number of technological and artistic advances. Massive ceremonial structures were constructed for public worship, burials began to be accompanied by elaborate grave gifts, and great care was accorded to the location and position of the body. Two methods were developed for the manufacture of cloth; textile designs of crabs, condors, and two-headed snakes were displayed on wearing apparel; and necklaces of bone and stone were common. Even the human skull was artificially deformed to achieve a particular aesthetic effect.

**Coca.** ***(Courtesy of the Fitz Hugh Ludlow Memorial Library)***

The first direct archaeological evidence of coca-leaf use actually predates this early agricultural epoch. Ceramic lime pots and figurines of coca chewers appear associated with the Valdiva culture on the coast of Ecuador and date back to 3000 B.C. There are also reports of coca use and even cultivation in some sites in the upper river valleys of Peru about 2000 B.C. Another site on the south-central coast, however, provides us with an even more interesting look at early coca use. The site, known as Asia I, includes a small village (50–100 persons) and a cemetery located in the lower Asia valley. The inhabitants of Asia I lived their lives in small rectangular rooms; when they died, they were packed into the high-density cemetery close to (and not unlike) the living compound. What makes Asia I so particularly interesting is that all the bodies found in the cemetery were buried in the same flexed position with the knees drawn up under the chin, all had artificially deformed skulls, and all were laid in the ground pointing west. They were wrapped up in large mats together with some personal possessions. Three thousand years later, when archaeologists carefully unwrapped these mats, they found mirrors, hairpins, wooden spears, clubs ornamented with sharks' teeth, snuff trays and tubes, and bags filled with coca and lime. The proximity of these goods suggests that the inhabitants of Asia I may have taken their coca in a fashion very similar to the way some people today take the plant's alkaloid, cocaine. The snuff trays and tubes at Asia I may have been used to fill the nose with coca powder as it is done today among the Indians of the Rio Miritiparana in Colombia.

Although the artifacts at Asia I represent our first direct evidence of man's association with coca, a number of secondary leads suggest even earlier dates. First, the lime found in association with the bags of coca leaves at Asia I indicates that, by 1300 B.C., users were aware that the leaves would yield their greatest effects when used in combination with lime. This means that considerable experimentation with the drug must have taken place prior to this time and suggests that the appearance of coca at Asia I represents only some of the first archaeological proof of an ongoing social practice that had been under way for quite some time.

Next, at least one author suggests that "the most ancient use of coca leaves in South American is its employment in various shamanistic practices and religious rituals."[1] In support of this theory, it can be said that even very ancient Peruvians had a sophisticated supernatural view of the world and depended on the occult powers of medicine men to cure disease.

Coca leaves would be a very important herbal addition to the repertoire of any medicine man or healer because of its anesthetic effect and particularly important to the shaman, who could alter his state of consciousness with the drug and, through the medium of the leaves, be able to feel closer to his god. Because the art of healing was basically religious and magical in nature, the roles of medicine man and shaman were very closely related in their primitive form.

Many ceremonial structures were built all over ancient Peru that served as the loci for most shamanistic activity. Perhaps the greatest of these was La Florida, located on the site of modern Lima. Excavations by the Museo Nacional de Antropologíca y Arqueología in 1962 demonstrate that La Florida represented the first great urban center in Peru that was built exclusively for worship. This complex of temples and altars was occupied by only a handful of priests and was deserted throughout most of the year. At certain times, however, the city was swollen with thousands of religious pilgrims.

We know that several types of plant substances were in use at La Florida for a variety of ceremonial and medicinal uses. Tobacco was used to a small extent for healing, *datura* was used for ritualistic purposes, and a snuff called *villca* was prepared from the ground seeds of the tree of the genus *Piptadenia* and used for its intoxicating effects. *Chicha,* the native beer, was consumed in great quantities at nearly all ceremonial gatherings and sometimes intoxication to the point of insensibility was expected of the celebrants.[2]

Coca was certainly among the imported plant stuff in use by the faithful at La Florida, yet its precise application at the ceremonies conducted there remains unclear. It may have been chewed by the worshiping masses to intensify the religious experience, or offered as a sacrifice. It is equally possible that coca may have been used to relieve the boredom that often accompanies lengthy ceremonial proceedings or chewed exclusively for its mild euphoric effect. We can be sure that whatever its use may have been, coca was a well-known and firmly established part of Peruvian life at least 2,000 years before the birth of Christ.

By about 1000 B.C., the old nomadic hunting and gathering way of life had completely disappeared in Peru, except for some isolated areas on the far south coast. In its place there was a gradual elaboration of sedentary traits that would later blossom into Peru's golden age. Population increased steadily but as yet there were no large cities. Pottery and maize were introduced from the north, and soon after, irrigation techniques were developed. It was not long before maize became the principal food crop and irrigation the rule for most important plants. Archaeological remains of the dog and llama appear for the first time, as well as the first examples of an American metallurgy.

As the Andean population increased and began to settle extensive areas on both sides of the mountain ranges, the establishment of sophisticated trade networks developed to accommodate the flow of goods and services across the continental divide. By this time, people were accustomed to bartering for specialty items from the eastern slopes and surely placed coca high on their priority list.

The period beginning about 200 B.C. and lasting to 600 A.D. was again a time of revolution and dramatic cultural change. For the first time, intensive warfare became a factor of everyday life. The peaceful villages that were once scattered along the Peruvian coast and highlands became, by comparison, large fortified cities that served as capitals for still larger political states. Archaeologists guess that by 600 A.D. the Peruvian population numbered four million. At least three cities had populations of 10,000 or more, each with a nucleus of public plazas and buildings, together with extensive residential districts that covered up to four square miles.

By this time, all known Peruvian food plants had been brought under cultivation. The major crops were maize, beans, peanuts, potatoes, sweet potatoes, yucca, chili peppers, pumpkins, papaya, pineapple, cotton, and coca. The inclusion of coca in this list means that by 100 A.D., the early Andeans had elevated the coca shrub to an extremely high position in their world. Of the dozen major crops just mentioned, coca is the only primary drug plant; if we do not count the cotton from which the Indians made their clothes and blankets, it is the only plant that was neither eaten nor depended upon for survival.

Another revolutionary change characteristic of this era came in the field of art. Textiles, pottery, metallurgy, and ceramic craftsmanship reached a very high level of excellence and it is generally agreed that this period produced many of the world's most distinguished works of craftsmanship. Two native cultures in particular displayed incredible quality in their art: Mochica figure painting or clay sculpture and Nazca embroidery have never been surpassed and rarely equaled in quality. These masterpieces have also provided us with an unparalleled look into the everyday life of ancient Peru. The surviving pieces capture the common man in the act of living out his daily life; some depict men harvesting crops, playing music, fishing, or taking coca.

Burial practices also changed during this period. By the earlier part of the first millenium A.D., bodies were wrapped in a cocoon of burial garments called *fardos*. Some *fardos* had great false heads with seashell eyes and hair made from sisal grass dyed black. Many wore bags or *chuspas* around their necks containing coca leaves. Others had gourds with maize, coca, and other plants inside. In the Canta mountains, some mummies were tied up in sacks made of llama skins. Between the leather bindings and the bodies there was a thick layer of coca leaves and wads of raw cotton. A tincture of coca is thought to have been one of the key ingredients used in the mummification process.[3]

Undoubtedly the greatest single burial find in Peru was made by Professor Julio Tello in 1929. At Parascus, just south of Pisco, Tello discovered 429 mummies stacked up in the corner of an ancient house site. Several of the mummies were wrapped in magnificent textiles with unmistakable Nazca designs. These embroidered cloths, now known as the Parascus textiles, are among the finest woven goods in the world. When Tello unwrapped and examined the mummies and their associated grave goods, he found many small bags filled with coca leaves. At another burial site located nearby on the south coast, more coca leaves were recovered along with gourds that were once filled with the intoxicating drink *chicha*.

**Peruvian priest with an artificially deformed skull makes coca offering at La Florida, 2000 years before the birth of Christ.** ***(Courtesy of the artist, Darrell Orwig)***

**Mochica pottery showing coca use, 200 A.D.** ***(Courtesy of the artist, Darrell Orwig)***

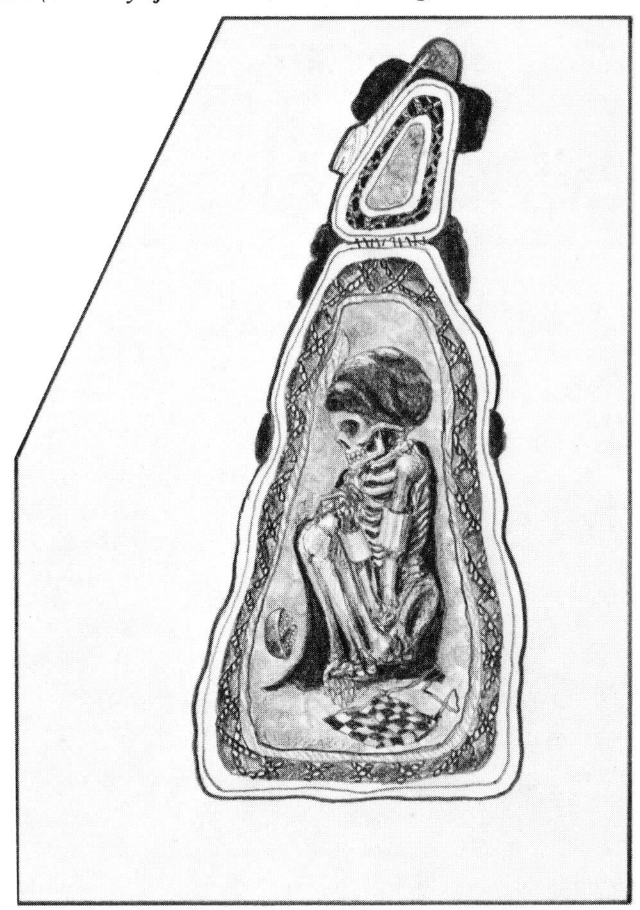

**Peruvian *fardo* or mummy. Coca leaves were put into the bags slung around the neck and were also placed among other goods inside the grave.** ***(Courtesy of the artist, Darrell Orwig)***

Thus, of the many sophisticated ecological relationships in ancient Peru, man's association with the coca plant was one of the most important. In support of this we know that coca figured significantly in the dramatic developmental period between 200 B.C. and 600 A.D., when Peruvian agriculture expanded to include all its important food crops. Next, the aesthetic revolution that produced so many pieces of high-quality art work in turn supplied artists with the means to display their ancient life-styles in action, and often this meant crafting clay or gold into representations of men with cheeks bulging from coca quids. In terms of coca's role in the development of established trade routes and population increase, it is enough to point out that there was a great number of people on one side of the Andes who desired the leaf but were unable to grow it, while another sizable group was settled on the eastern slopes growing coca in surplus and willing to provide it to others. By this time, thousands of Indians were aware that coca had the power to overcome pain, fatigue, and hunger, the power to enhance spiritually any religious occasion, and the power to please.

The coca shrub rooted itself in the people's leisure time, their religion, and their work. It retained a reputation as an all-purpose cure in the local pharmacopoeia and was so omnipresent in ancient Peru that people were as often as not buried with it. In view of all this evidence, it is not unreasonable to say that in this formative period, the ancient Peruvians embraced the coca plant and wove it directly into their social fiber.

For the next 1,000 years, little is known about the movement and use of the coca plant. At least two great empires emerged and collapsed during this period, but the names and languages of these people and their customs are lost forever; no written records or oral legends have survived into modern times. We may assume, however, that coca use continued on as something of a tradition in this area and survived to experience periods of popular interest as well as times of disinterested neglect.

**Mochica vase depicting a priest with a coca bag.** ***(Courtesy of the artist, Darrell Orwig)***

As we shall soon see, the tribal group known as Inca appears to have continued this pattern until the early 1400s, when a king named Pachacuti began an effort to make the coca plant the single most important item in his world.

# 2

# The Incas: Cocaine Adaptations in the First Millennium A.D.

Twenty miles southeast of Cuzco is the place once known to all Incas as Paritambo, or the Origin Lodge. Here, from caves in a sacred mountain, came four legendary brothers and their sisters/wives, all children of the sun god, Inti. These heroic characters were destined by their father to rule the world by conquest and administer their father's truths to all the people with whom they came in contact. In their possession was a golden staff that functioned as a kind of divining rod in the selection of the place where the new empire was to have its capital city. After one of the brothers was murdered, another turned to stone, and a third metamorphosed into a large birdlike creature, the surviving brother, Manco Capac, and the remaining women wandered on to the abandoned site called Huaynapata, where the golden staff plunged into the ground and disappeared forever. This was the sign that they had reached "Cuzco," the place Inti said the Incas would finally settle.

This Inca origin legend has associated its characters with real beings, places, and times. Although the story is mythological, it is generally agreed that the first Inca sovereign was a man by the name of Manco Capac, and that he established his rule sometime around 1000 A.D. at a place between the Huatany and the Tullumayo rivers. At this time the land was occupied by Indians of the Hualla, Alcaviza, and Sauasiray tribes; in fact, the name "Cuzco" was the Hualla name for the entire area.[1] The beginning of the eleventh century was a time of ethnic unrest in Peru, and Manco Capac emerged as one of the most powerful war leaders, or *sinchi,* elected by his people to insure the survival of the group. As his band increased in number, Manco Capac's sphere of influence enlarged until it incorporated many neighboring tribes by either war or treaty.

In the early days of Manco Capac, coca was already a well-known plant substance. Its powers were legendary and the known history of its use featured many mysterious implications. Because of this, coca became the object of numerous legends that were designed to explain the plant's appearance and fix its place in the minds of the Inca people.

The most popular of the Inca myths concerning the leaf tells of a beautiful woman who had to be put to death because of her sexual desires. The legend says she was cut into pieces and from these parts came the shrub called *mamacoca.*[2] Her assassins then carried the plant in small bags that were forbidden to be opened until after they had had intercourse with a woman.[3]

Another story describes the terrible storm of Khuno, the god of snow and tempest. When the Indians of the *altiplano* crossed the mountains and settled in the high valleys of the *montaña,* they found it necessary to set fire to parts of the jungle to make it fit for cultivation. Khuno was angered by the clearing fires beacuse they sent up smoke that polluted the peaks of Illimani and Illampu. In retaliation, he sent down a fierce storm that destroyed everything and sent the people running for their lives. Rivers became swollen and the swift currents swept away roads and opened huge ravines. When the storm passed, the farmers returned from the caves where they had taken shelter and found complete desolation. They began to search for food and soon came upon an unknown shrub with brilliant green leaves. They gathered the leaves and put them in their mouths and immediately were suffused with a sense of well-being. Because of this plant, they were able to escape the devastation of the storm that ravaged the land and they lived to reveal the secret of coca to their descendants.[4]

A third myth is set in a time of great famine. Manco Capac looked down on the world from heaven and wept bitter tears for his suffering children; his tears amounted to a downpour and saturated the ground below. Later, all the tears evaporated, except for the ones that settled on coca plants. Manco then sent a blazing red comet to earth as an omen to his people. It landed in front of the palace of a king named Montana. When the king rushed out and approached the comet, it transformed itself into a flaming coca leaf. After the fire subsided, Montana kissed the holy object and tasted the bitterness of Manco's tears. Montana understood that he should give the plant to the Incas to defeat hunger and fatigue. With the famine behind them and the coca leaf giving them power, enthusiasm, and energy, the Incas prospered.[5]

Another legend speaks of the days called Purun Pacha, a time of hostility and angry spirits. Coca was considered an evil and poisonous plant until a goddess with a child came from heaven and asked the plant for shelter. She was able to rest beneath the plant and hang her child's clothes on its branches. From that time on, coca was favorably received by the Indians and was considered a source of strength.[6]

Undoubtedly the most intriguing legend attributes the spread of the plant to the fascination held for a conquereror by the bright eyes of a beautiful native princess who loved him to death.[7] A. A. Moll, the source of this story, unfortunately does not provide the details.

**Golden Inca figurine with bulging cheek. Photo by Loren McIntyre.** ***(Courtesy of Loren McIntyre)***

Other indications place coca in association with many aspects of the early Inca period. If we trace the word *coca* to its beginning, we find that it comes from the local Amarya term *khoka,* meaning simply "the tree," suggesting that it was a cornerstone in their botanical world, their divine plant, or perhaps, the tree par excellence.[8]

Inca genealogies reveal that the wives of the second and fourth Inca named themselves after the legendary *mamacoca.* Later, many Indian girls took the name to mean a thing of power.

Around 1150 A.D., the third Inca monarch, Lloque Yupanqui, was so interested in the leaf that he led a large army into the inhospitable *montaña* in search of it. Sometime during this campaign he established what is thought to be the first great Inca coca plantation at a place called Havisca. The historian Garcilasco de la Vega Inca (who was the offspring of a Spanish soldier and an Inca princess) later came to inherit this *cocal* 350 years later and reported that it was still in full-scale production.[9] The highland soldiers of Yupanqui's army found the steaming jungles of the eastern slopes difficult to endure. Mosquitoes hovered around their heads while



SAINT PETER'S COLLEGE LIBRARY
JERSEY CITY, NEW JERSEY 07306

they marched and centipedes bit their feet. Those who did not fall victim to a variety of tropical diseases were often halted by five-foot poison-tipped arrows that were fired from ambush by men who hunted human heads as trophies. Diet was also a problem and the men complained that there was never enough parrot meat or roots to keep them satisfied. Some combination of these unsavory elements dampened Yupanqui's coca curiosity to the point where he elected to return to Cuzco and disassociate himself and his people from the plant altogether.

Inca interest in the coca leaf seems to have endured a 150-year impasse until a tall, smiling man named Inca Rocca took command of a large army and marched south and to the sea to conquer the Nazca nation. These coastline people surrendered to the Inca forces without much of a struggle and, as a result, young Prince Rocca found himself with enough time to court the more gentle Nazca customs and pastimes. Of all these, he became particularly fond of chewing a small green leaf that the Nazca used to dull hunger and cheer the spirits.[10] The Nazcas had obviously been enjoying coca as a trade item from the *montaña* and had long ago incorporated it into their society.

When Rocca returned to Cuzco, his father saw him chewing leaves and asked him if he had acquired this habit during his recent campaign. Rocca admitted that he had and offered his father some leaves from a finely woven bag. The Inca refused, saying he already knew about coca from his grandfather and had even experimented with it. He warned his son not to recommend its use to the people. Unfortunately for the old king, his warning had come too late, for an estimated 10,000 soldiers from the Nazca campaign had come to use and enjoy the leaf and spread its fame throughout Cuzco.[11]

Shortly thereafter, the old man became ill and died, and Rocca became the sixth Inca. After the customary two-year mourning period, he took up arms again, this time striking north and conquering the Chancas. Rocca's empire now stretched north, south, and west for some distance; only the eastern home of his beloved coca remained to be brought under Inca rule. To complete the job, Rocca assigned the task of conquest to one of his captains named Apo Camac ("the tiger"). Camac's men suffered the same hardships as their great-grandfathers had the last time a highland king ordered a military force across the great divide. After several frustrating months in the jungle, they were forced to return to Cuzco with only a small part of the *montaña* in Inca hands. Apo Camac never returned, but instead chose to remain with and marry into the savage headhunting tribe known as the Chuncho.

Apparently, the campaign netted enough land to supply Rocca and his people with the much-desired leaf, for he is sometimes remembered as the man who first brought coca to the Incas.[12] While he is certainly not the first to have introduced the plant to these people, he very well may have been the first to mount a large, successful campaign against the people of the coca-growing regions with the express purpose of securing their lands for exclusive Inca use.

The reintroduction of coca to the Incas had not only won a symbolic title for a young king, but it also set the stage for a later and more powerful king named Pachacuti to manipulate by exclusion and elevate the "tree par excellence" to one of the most valued commodities on earth.

The first seven Inca rulers contented themselves with small raids on neighboring tribes, for purposes of either consolidation or revenge. It was not until the ascension of the eighth Inca, named Viracocha, that the still tiny state began to consider the possibility of an empire based on

**Inca Rocca is said to be the man who first brought coca to the Incas.** ***(Courtesy of E. Riou)***

a full-scale invasion and tribute. Viracocha made several tours of the area prior to his planned invasions. During his absence, it is said, his lonely wife consoled herself with dwarfs and hunchbacks and made use of coca to excess.[13]

As a result of his journeys, the Inca knew that invasion would mean occupation. If he wished to secure these new provinces permanently and make them pay tribute, he would have to introduce Inca rules and back them up with Inca enforcement. For the local cooperation required in any successful operation of this kind, Viracocha realized he must employ powerful control measures over village people; consequently, his methods for social control were designed to minimize local power and, at the same time, maximize dependence on the Inca.

It was decided that one of the first things to be done after the acquisition of any new territory was to send the most dissident members of the newly conquered people to an older and more established section of the Inca empire. These people were in turn replaced by a like number of Incas who were known as *mitimaes.*[14] Sometimes entire nations were moved according to Inca wishes. The Lupacas people of the province of Chucito near the shores of Lake Titicaca, for example, were resettled in a part of the newly conquered *montaña* and were forced to provide coca for the Incas. In return, they were sent *quinoa* and *chuno,* which was their regular diet.[15] Control was also exercised through incorporation, whereby chiefs of local tribes were accepted in accordance with their rank and were given Inca women as wives. Women and coca came to share an unlikely role under the Inca system. They were both kept as state treasures and were used as gifts or rewards to cement relations with conquered chiefs.

Young men of fighting age were often impressed into various Inca armies and sent off to fight wars thousands of miles from their homes. This military conscription helped remove the bulk of young male resistance from the new territory and at the same time increased the strength of the Inca forces in the field.

In addition to these measures, the Inca tongue, Quechua, was introduced as the official language of the empire and was used to conduct all state business. In distant parts of the empire, where Quechua was totally unknown before the invasion of Inca troops, the language conversion was slow and must have complicated their designs on complete communication control.

Although every attempt was made to uproot and destroy local traditions and replace them with the Inca system, these various methods of social control were received with only varying degrees of success. Recalcitrant tribes often held fast to their own beliefs about things and thus weakened the power of an Inca governor representing the wishes of his god/king.

Time and time again the Incas were faced with the task of trying to subjugate newly conquered people by attempting to dissolve local traditions and replace them with the Inca way. The problems involved with this system were numerous as various groups resisted Inca control methods and tenaciously held on to their ethnic identity.

The Incas needed to locate a common denominator for control purposes or, in other words, a single culture trait that was shared by many of the societies under their rule. If they could completely monopolize a trade substance desired by a large percentage of the people they conquered, the Incas could at once bring about a degree of social control by placing restrictions on the distribution of this substance. This type of selective embargo militated in favor of rapid subjugation and facilitated Inca efforts to bind new people to the state in Cuzco. The Incas may have found this substance in the coca plant.

By the fourteenth century A.D., the coca plant was undoubtedly in use over a large area extending from the Amazon Basin across the Andes and from Chile to Colombia, where the habit of chewing leaves with the addition of lime has been called an "old and widespread practice."[16]

Although the preferred leaf is grown exclusively on the eastern slopes of the Andes, in the *montaña* or *yungas* region, we know that coca was proliferated throughout the northern half of the continent by the many sophisticated cultures that preceded the Incas.[17] It moved along established trade routes for centuries and was surely a familiar and valuable trade item for millions of South American Indians.

Because of its powers, coca was always a much-sought-after plant stuff that was thoroughly mixed up with the magico-religious healing syndrome associated with the Inca and so many other early peoples. Its additional value as a mild intoxicant assured coca a place alongside *chicha* in most Andean settled communities that could lay hands on it. The management of the coca leaf in each of these communities is unknown; we can say, however, that its administration and restriction were probably as varied as the customs of the many tribes that used it. Whatever its specific role may have been, coca was above all a preferred substance in use by a great number of Andean societies and ripe for Inca seizure and control.

# 3

# The Most Valuable Substance in the World

The empire-building notion actualized by Viracocha was thrust into high gear by his son, Pachacuti, whose name means "one who changes the world." Under Pachacuti, Inca campaigns were no longer mere raids; large organized armies struck out in all directions from Cuzco intent upon permanent conquest. Pachacuti immediately began to refashion the Andean world on three fronts simultaneously: imperial expansion, administrative integration, and ceremonial sanctification.[1] His expansion was so sudden and successful that in the space of fifty years the Inca realm stretched from northern Equador to central Chile; it encompassed 350,000 square miles of territory and millions of Indians representing hundreds of different cultures.

Pachacuti soon realized that the administrative integration and ceremonial sanctification he desired could be better realized through a unique blending of these two aspirations and the coca leaf. In order to seize more control through integration on the administrative level, Pachacuti turned his armies east toward the *montaña* in a move designed to take control of some gold mines and, more importantly, the choice coca plantations. Gold, the teardrops of Inti, was held in high esteem by the Incas, but coca was more valuable than gold to the majority of the Andean peoples. Pachacuti realized that a successful military campaign in key coca-growing areas would place in his possession the production centers for possibly the most prized commodity in his world. In his hands the shrub would be divine and its distribution as a divine plant would be determined only by the Inca. This maneuver could be called the politics of relative deprivation and was very much in keeping with Pachacuti's tireless quest for power and authority.

In the last days of the Chanca Wars, Pachacuti led his armies in an attack against the stronghold known as Ollantaytambo. This fortress commanded two of the possible three routes into the Vilcabamba and was the door to the *montaña* and coca.[2] Interestingly, Inca Rocca, the man who is said to have introduced the leaf to the Incas, had razed Ollantaytambo to the ground many years before in his attempt to obtain the leaf of the eastern slopes. Now the same city was under attack once again, this time by Pachacuti, who had the same designs as his great-grandfather, only on a much grander scale. The huge empire he envisioned would require great warehouses full of coca that would have to be constantly supplied. To realize this end, his large army moved through the pass of Panticolla behind Ollantaytambo and headed for Victos, the central pueblo of Vilcabamba. Pachacuti accepted the surrender of Victos without losing a man and immediately began to organize the district in a way that would more directly benefit the needs of his developing empire.

Gangs of Inca road builders were sent to improve the trails to highways while engineers and herbalists taught the people to mine gold and grow the sacred coca leaf.[3] The workers in the

**Pachacuti leads a large army to the coca regions.** ***(Courtesy of the artist, Darrell Orwig)***

plantations were divided into *ayllus,* or groups of ten; young men from sixteen to twenty years of age were set apart for light work and were known as *coca-pallac,* or coca pickers.[4] Inca construction workers erected numerous buildings at the borders of the district, many of which were storehouses to hold the gold and coca that was to be exacted as tribute.

Later, Pachacuti's army pressed into the Carabaya region. They marched through the passes of the Nudo de Quenamari and once again penetrated the inhospitable *montaña* in pursuit of its riches.[5] Inca specialists poured into the area and another road was built to facilitate the removal of goods.

Although some species of coca can be found along the eastern slopes of the Andes all the way from the Caribbean Sea to the Straits of Magellan, the very heart of the coca region is the Peruvian *montaña* and the Bolivian *yungas.*[6] It is from these regions that the finest variety of coca is grown and it is here that Pachacuti concentrated his efforts. He knew that several large *cocals* in this region, if organized properly, could supply the Incas with enough good leaf to satisfy his expanding empire. In support of this, we know that at one time Paucartambo and several other Indian towns along the Huanuco Valley were credited with supplying coca for all of Peru.[7]

Despite the plant's wide range, the Incas now controlled the finest growing areas and possessed the organizational powers and military muscle to do with coca what no one else had ever done before: they had the power to circumscribe. The Incas were perfectly equipped to produce, store, move, and remove large amounts of coca and apparently did so as they saw fit. With the Inca armies in virtual control of most trade routes and nearly all major highways, Pachacuti could regulate what goods passed from one district to the next.[8] It was now a small matter to determine who in Pachacuti's empire would have access to a commodity such as coca.

It was at this point that coca became a re-

**Coca terraces in the *montaña*. *(From* Bulletin on Narcotics *4, no. 2, 1953 )***

stricted substance throughout the Andean world. For the people of Cuzco, use of the leaf had been confined to the royal family for some time already, probably since the maturation of the religious system or about the time of the establishment of the origin myth. As the theocratic Inca state developed into an imperialistic force, more and more tribes were faced with the acceptance of the Inca rules. Now "it was unlawful for any of the local people to use coca without permission from the [Inca] governor."[9] For the first time in historical record, the coca plant was used to serve a political end. By lording over the distribution of their divine plant, the Incas under Pachacuti assured themselves of an additional degree of social control on at least one important level: the plant that was used for so many reasons by so many people now had to be obtained through the Inca, as all foreign trade was his state monopoly.[10]

The advantages of being in this position were obvious, and Pachacuti was the type of man to take advantage. Under his rule, an expansive yet cohesive state came into being. He made the capital city of Cuzco a large metropolitan mecca for the many people in his nation and filled it with large public structures of superior workmanship and design. Pachacuti is responsible for the reorganization of the Inca state; he is the real father of modern Inca ritual and theology and, as a conqueror, he is the greatest American in history.

It was the Inca state at its greatest moment, under Pachacuti, that was related to the first Spanish chroniclers, for there were still some Indians alive at the time of conquest who remembered seeing Pachacuti and who had lived under his glorious rule. From these firsthand accounts, we can begin to see the extensive role that coca played in the lives of these South American Indians at the close of the fifteenth century.

Coca was a very formal and important institution during the height of the Inca rule. The term *huaca* represents a basic concept in the understanding of the Inca relationship with the coca plant. Like the Sanskrit term *prajna* or the Polynesian *mana, huaca* was the special power resident in a place, thing, or person. A mummy, a stone, or a mountain could be *huaca,* or it could be the spirit that enables some craftsmen to do exceptionally fine work or a fisherman to catch many fish. For the Inca, coca was *huaca.*[11]

Although it was not actually worshiped, coca was considered divine because it was a means of force and strength for the rulers of Cuzco.[12] It is said that Pachacuti had a garden outside his palace in Cuzco called Coricancha with representations of coca leaves casted out of pure gold.[13]

While coca was *huaca* and ordinarily for the exclusive use of those with royal blood, a number of exceptions were made to some regular citizens on the basis of their occupation. Official state messengers call *Chasquis,* whose job it was to pass information and small parcels throughout the empire, were permitted recourse to coca.[14] These messengers were selected sprinters who would run from station to station carrying their messages pony-express style. By this method and undoubtedly with the aid of coca, messages moved through the Inca state at an incredible rate of 150 miles a day. The Inca was able to eat salt-water fish in Cuzco that was caught in the Pacific only the day before. It is important to realize that the run from the seashore to the Inca dining room was 300 miles, all uphill.

Men called *yaravecs* were allowed to use coca. The *yaravecs* were court orators and were, in effect, what the Inca used for the printed word. They were living history books and were required to memorize, among other things, even the finest points of Inca genealogical happenings, both past and present. They were assisted in these phenomenal memory displays by knotted string devices called *quipus* and, again, the coca leaf.[15] Special priests known as *virapiricues* made daily sacrifices of coca in the Temple of the Sun and also used the leaf to foretell the future by burning them and interpreting the smoke. Coca was in fact used as a sacrificial agent by priests and some others in virtually all ceremonies throughout the year, and it was generally agreed that any important affair attempted without an offering of coca could not prosper.[16]

Coca was used in the initiation rites of the young nobles in the festival called *uma-raymi.* At the end of a very arduous series of events, a foot race took place to determine who was the strongest and fastest of the contestants. Young women lined the route the runners used, encouraging and stimulating them by saying, "Come quickly, youths, for we are waiting," and giving them *chicha* and coca.[17] Those who survived the ordeal of the *uma-raymi* were awarded a *chuspa,* or woven bag filled with coca leaves that was symbolic of the supposed strength and manliness of the young Inca noble. The *chuspa* was also an essential part of the emperor's wardrobe and was considered part of his impressive collection of power symbols.

The leaf was also a reward for those lucky individuals who met with imperial favor. Its use may be compared to the laurel leaves a Roman general received after a particularly successful campaign. Nobles of defeated tribes were also sometimes granted permission to possess and use the leaves.

Coca plantations served the Inca jurisprudence system in a most direct way. Convicted thieves were first required to compensate their victims and then were sentenced to hard labor in the state's *cocals.* Because work in the jungle was so unpleasant and unhealthy, criminals were usually allowed to apply for parole after a relatively short time.

The coca leaf also figured in the practice of human sacrifice. The Incas, in fact, inherited this gruesome custom, for it is known that it was a common practice among the many tribes that preceded them in that region. The Inca historian Garcilasco tried to claim that the "children of the sun" did not practice human sacrifice, but he was wrong. It can be said, however, that the ceremony was suppressed under the Incas and that human beings were killed and offered only with state sanction and only on special occasions. Unlike the Aztecs, who tore beating hearts from living men, there is some evidence that the Incas revered their sacrificial victims to the point of attempting to anesthetize them with coca to help them endure the ordeal. There is also evidence of women being buried alive next to some mummies. It is suspected that these ladies represented the least fortunate of the mummies' widows, who were sent to the grave along with their husband. When this practice occurred, it is said that the unlucky woman was first drugged with coca powder.[18]

There have also been a number of Inca sacrificial victims found frozen to death on the top of mountain peaks in the Andes.[19] An Inca boy about nine years old was discovered on the top of 17,000-foot Mount Plomo in northern Chile and appears to have been given coca to help numb him from the cold and ease the pain of his death. With him in his rock grave were bags containing his baby teeth, hair combings, and fingernail parings. There were also seashells, a tiny gold llama, and two large bags filled with coca leaves.

Coca also played a large and important role in Inca medical practices. Inca physicians believed coca was an aid to longevity and in fact certain areas of the Inca domain do boast long histories of citizens who have survived well past their hundredth year.

Like Babylon and Egypt, priestly medicine was common to Peru. One reason for this is that the people all believed that disease was caused by "sins" against the religious or political system. Moreover, the Incas were given to the notion that certain rituals of atonement could be performed to relieve physical suffering and cure disease. When these important ceremonies were undertaken, the presence of the coca leaf was absolutely essential to the Indians by helping them blend the religious and medicinal aspects of the occasion into a single dramatic event.

Inca medicine men and healers used coca diagnostically as well as therapeutically. Some diagnosticians divined the cause of a particular unknown disease by referring to the positioning of coca leaves sprinkled on the ground or by the route of coca juice spit on the hand. Therapeutically, holy coca was the leading drug in the Inca materia medica and appears to have been employed as a local anesthetic as well as a stimulant.[20]

The anesthetic properties of coca were enjoyed by a great many Andean surgery patients. The paleopathologist R. L. Moodie reports, "I believe it to be correct to state that no primitive or ancient race of people anywhere in the world had developed such a field of surgical knowledge as the pre-Columbian Peruvians. Their surgical attempts are truly amazing and include amputations, excisions, trephining, bandaging, bone transplants, cauterizations and other less evident procedures."[21] In support of this, we know that the number of trephined skulls attains an all time peak in the Peruvian collections.[22]

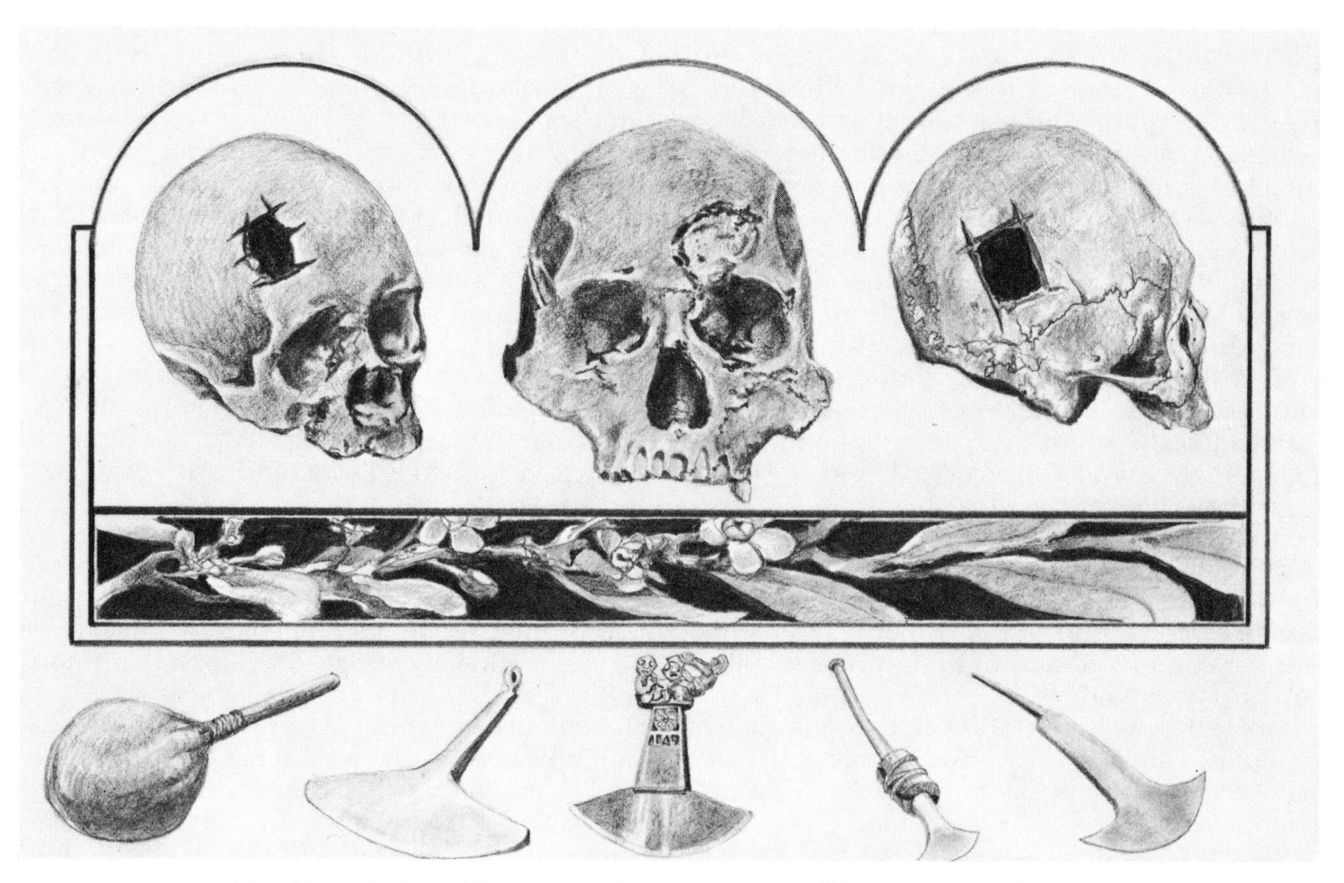

**Trephined skulls and Inca surgical tools.** ***(Courtesy of the artist, Darrell Orwig)***

Some skulls bear no less than four trephination holes and many show signs of new, healthy bone grown in well after the initial cuts were made. Trephining was generally performed to relieve the compression in punctured and comminuted skull fractures, a type of injury that is commonplace where warfare with spiked clubs and sling stones is an almost everyday affair. Coca-rich saliva appears to have been applied directly to the wound as a local anesthetic and there is some evidence that it was also administered as a suppository. In the Ollachea Valley there are several tombs containing unusually large numbers of trephined skulls in association with syringes used for enemas.[23] This suggests that some type of coca solution may have been introduced directly into the large intestine to take advantage of the rapid absorption that takes place there.

All Incas considered coca to be a powerful aphrodisiac and it was prescribed and administered accordingly. The Inca Venus is represented holding a spray of coca in her hand.

The overall importance of coca in the everyday routine and ritual of the Incas is so extensive that it is impossible to calculate. Inca life and coca are mixed up so thoroughly—economically, militarily, medicinally, spiritually, and sexually—that it would be difficult to imagine one without the other.

Although millions had taken the leaf before the arrival of the Incas, it was the people of Manco Capac who took the leaf as no one ever had before or since.

# 4

# A Small Band of Curious Men: Pizarro Seizes Peru and Coca

Pachacuti, by now an old man of eighty, abdicated his throne to his favorite son, Topa Inca, in the year 1471. At age eighteen, the young prince had marched against the powerful kingdom of Chimor and later had consolidated many coastal states into his empire. Though his victories were costly, Topa Inca succeeded in demonstrating to his father that he intended to expand and rule the empire in a fashion that would both imitate and honor the old king. Accordingly, he was in the Sierra fighting when Pachacuti died in 1473.

While Topa Inca hoped to mirror his father in all ways, he could never exercise the personal leadership of the great king, and the strict control that built and maintained the Inca empire began to weaken. After defeating Chimor, Topa Inca stopped fighting and divided his time between elaborate tours of the empire and planning extravagant building projects in the cities. He is known to have fathered close to 100 children and is also remembered as having popularized the widespread use of coca among the nobility.[1] Conspicuous consumption of the leaf by the fashion-minded, privileged class also had its effect on the masses. As Topa Inca's royal entourage drifted from one district to the next, their bulging cheeks and enthusiastic references to coca served to inspire a renewal of popular interest about the plant.

Shortly before the Lupaca rebellion took place near the shores of Lake Titicaca, Topa Inca had occasion to march a large army across the Andes and deep into the eastern forests of the Antisuyu; to facilitate travel through the dense jungle, he ordered the construction of an enormous fleet of canoes. After much time and energy had been expended toward the completion of these vessels, the Inca army sailed downriver until they reached the base camp of the barbarian invaders. Once there, they exercised what has been termed "an excessive amount of force to crush a relatively tiny force of Amazonian raiders."[2] Several authors have questioned the great military efforts put forth by the Incas to check the occasional raids carried out in the distant eastern section of the empire and the substantial economic efforts directed at the construction and maintenance of vast fortresses that were designed to insure the safety of what was, in fact, only a jungle. The alpine Incas enjoyed the luxury of having the mighty Andes to protect their eastern flank and were never at any time threatened by an invasion of Amazonian people. For justification of these military and economic acts, one need only realize that the endangered area in question contained the state's coca plantations and that at the time they were being worked by a criminally dissatisfied element of the Inca population who were sent there to serve out their sentences.[3]

Topa Inca and his army had barely secured

**Topa Inca.** ***(Courtesy of E. Riou)***

**Huayna Capac.** ***(Courtesy of E. Riou)***

the *cocals* before they were informed of the Lupaca rebellion near Lake Titicaca. Without hesitation, the Inca army mobilized and began a forced march that originated in the tropical rain forest and ended at a 12,000-foot mountain plateau next to the great highland lake. Much has been said about the training, discipline, and especially the remarkable endurance of the Inca troops that played a part in this logistical miracle. Surely Topa Inca took advantage of the stimulating properties of the leaf he labored so hard to popularize and protect by issuing it in fresh, whole lots to his expeditionary forces departing from the *montaña*.

In the same year that Christopher Columbus reported to Queen Isabella on the New World discoveries he thought to be on the fringes of Cathay, Topa Inca died in a retreat near Cuzco. He left a legacy of sixty-two sons and thirty daughters behind, among them a prince named Huayna Capac, who was to become the eleventh Inca and the inheritor of a world of changes.

Unaware of what lay just ahead, the boy-king Huayna Capac launched the Inca state into what has been called "its most elegant reign."[4] He ordered a year of celebration and spent most of his time moving from one festival to the next. He managed to produce at least three heirs to the throne by the time he was fifteen, and acquired a reputation for his abilities with *chicha*. Early in his reign coca-leaf use had grown to become "a common minor vice among the nobility and by this time was also in widespread use among the commoners."[5] Although it is generally believed that coca remained a restricted substance until the Spaniards arrived and destroyed the social system, Garcilasco correctly reports that by the time of the Spanish conquest, the coca leaf was already in the process of becoming available to everyone.[6]

With the prohibition essentially lifted, popularity and accessibility reached an all-time high. Paved Inca roads now served virtually all the coca terraces of the *Antisuyu* and the drying patios of the lower Urubamba.

Under Huayna Capac, large quantities of coca were free to travel the Inca highways into the far-flung reaches of the empire, and the leaf was once again in the public domain. Scarcely thirty years after Pachacuti's death, the commodity he fought wars to restrict was now in use by more people than ever before in history. Even more ironic is the fact that the Europeans, who knew nothing about the leaf at the height of its popularity, would soon realize coca's importance to the Indians and, like Pachacuti, would manipulate it against them to advance their own ends.

This complex process began in early September of 1499 when an amateur navigator from Florence named Amerigo Vespucci became the first white man to enter the history of coca. Vespucci abandoned the routine life of a banker's

**Drying patio with mound of fresh coca leaves. *(From* Bulletin on Narcotics *4, no. 2, 1953).***

assistant to join an expedition bound for South America with Alonso de Ojeda. After reaching the Amazon River, the coast of Paria, Venezuela, and Magdalena, he returned to Spain and later took a position with the Portuguese crown. Many of Vespucci's subsequent writings, such as *Novus Mundus,* are proven fabrications and it is at best a dubious honor that both continents of the New World accidentally became known by a term that was a misspelling of this man's first name, (Vespucci's Christian title somehow appeared on the Waldseemuller map of 1507 and afterwards was generally mistaken as the name assigned to the new land). In spite of all this adverse information, Amerigo Vespucci did give the world its first account of coca. In a letter (which seems to be genuine) to his friend Piero Soderini, he tells of his coca observations on Isle du Margarita. "The customs and manners of the tribe are of this sort. In looks and behavior they were very repulsive and each had his cheeks bulging with a certain green herb which they chewed like cattle, so that they could hardly speak, and each carried from his neck two dried gourds, one of which was full of the very herb he kept in his mouth, the other full of a certan white flour-like powdered chalk. Frequently each put a small powdered stick (which had been moistened and chewed in his mouth) into the gourd filled with flour. Each they drew it forth and put it both sides on his cheeks thus mixing the flour with the herb their mouths contained. This they did frequently and a little at a time, and marvelling at such a thing, we could not guess the secret nor for what purpose they did so."[7]

A dozen years after Vespucci's comments, Vasco Nuñez de Balboa, still in the Caribbean, first heard tales of a golden mountain to the South. At the same time, an illiterate conquistador in Panama named Francisco Pizarro listened intently to the old Indian stories of "El Dorado, the City of Gold" that had been passed along the trade routes for decades. The emperor Huayna Capac was then about 23 years old and was concerned primarily with the outcome of the Carangi uprising in northern Equador. A dozen years later, while the Inca celebrated his long-fought victory in Quito, Pizarro, Father de Luque, and a man know as Almagro met secretly in a small room in Panama and divided the spoils of a land and empire they had never seen.

In 1522, no Inca had ever even heard of a white man, yet white Europeans were affecting the Inca nation in a most dramatic fashion. During the victory celebrations in Quito, Huayna Capac received word that renegade Chiriguanas Indians had invaded a Bolivian province; he did not realize that the leader of this group was a bearded Spaniard by the name of Alexio Garcia who had washed ashore in Brazil some years before. Garcia came to lead the raiding party in search of a place known to him as Caracaras, where there was supposedly a huge silver mountain and great riches.

While this minor disturbance was taking place, a much more serious threat to the Incas was in progress. A horrible disease similar to both bubonic plague and typhus was killing off many Inca soldiers. The sickness was brought by the Spanish armada of Pedrarais in 1514 and had slowly moved through the mangrove swamps of Colombia until it struck the Inca ranks at the close of the Carangi wars.

These pieces of bad news were soon compounded when runners informed the Inca of

**Amerigo Vespucci was the first man to write about coca use. He said, "and each had his cheeks bulging with a certain green herb which they chewed like cattle, so that they could hardly speak. . . . This they did frequently and a little at a time, and marvelling at such a thing, we could not guess the secret nor for what purpose they did so."** ***(Courtesy of F. J. Pohl)***

large sailing vessels with strange bearded men in the coastal waters.

The combination of these disturbing reports prompted Huayna Capac to retreat to Tumibamba. Upon his arrival, messengers told him of the mounting death toll in Cuzco because of the plague. Among the dead were many nobles closely related to the Inca Chief, including his sister, Mama Coca.

Fearing that all these events were in some way related, Huayna Capac consulted oracles about the bearded strangers—were they gods or were they men? The prophecy was to be an important one, for all throughout the empire there were rumors that the twelfth Inca would be the last.

By now the disease had killed 200,000 citizens in Equador and sacrifices and penances were undertaken by everyone, including the king. Huayna Capac retired to his quarters to fast and contemplate the arrival of the bearded men, who were still hovering off the coast. Later, the emperor is said to have dreamed that three dwarfs came to visit him and had left a mysterious box that was supposed to have contained the disease. The nightmare unnerved the king and seemed to bring on a fever. A day later he took a turn for the worse and Huayna Capac predicted he was going to die. Certain in the knowledge of his immediate death, he called the priests and other Inca officials to deliver his last words regarding the bearded men. He stated that they were the people of Viracocha and that they were responsible for the sickness. He was apprehensive of them and warned the Incas to placate them if they must.

After the meeting, the condition of the Inca grew considerably worse and special priests were called in to try to save him. They blew tobacco smoke in the air to rid it of any evil spirits and sprinkled flour made from black maize on the floor. They arranged two pottery fixtures and built fires in them with pieces of wood dipped in llama fat. When the coals from the fire were red hot, the medicine men folded their legs under them and began to chew large quids of coca. They looked into the fire and attempted to interpret the flames and smoke. They said, "Tell us from where comes this disease? What is it? And who has brought this to our Inca?" By the following evening, the eleventh Inca lay dead in the northernmost fringe of his soon-to-be conquered empire. His untimely death added to an already confused situation, for he passed away leaving the heir to the Inca throne in doubt.

When Huayna Capac died in 1525, he left two prominent sons: Huascar, son of his sister/wife and legal heir; and Atahuallpa, son of a secondary wife and the king's favorite. Both men assembled large armies soon after their father's death and began a protracted war to determine who would be the twelfth Inca ruler. While the extensive preparations were being made for this civil war, Francisco Pizarro and his ship of Spanish adventurers skirted the coastal waters of Peru. With the spotlight of attention focused elsewhere, the Spanish were able to make several stops on the mainland and received gifts of gold, women, and llamas. The shrewd Pizarro soon calculated the extent of Inca wealth and, as if to allow the war of the two brothers to play itself out uninterrupted, he sailed back to Panama and later Spain to regroup and consolidate his plan of attack.

His return to Spain excited the interest of Charles V. In July of 1529, Pizarro was given permission to return to Peru and seize it in the name of the king of Spain. He and his party left Seville in January of 1530.

By April of 1532, Pizarro and his men had crossed the Atlantic and landed on the north coast of the Inca empire near Tumbez. Here they began a series of raids and skirmishes, looting and killing their way south towards "El Dorado, the City of Gold."

As if according to a novelist's plot, the second arrival came almost exactly at the time of the defeat of Huascar at the hands of Atahuallpa. Self-assured and secure after the kidnaping and defeat of his brother, Atahuallpa retired to a warm-spring retreat outside the city of Cajamarca. There he lightly entertained the reports of Pizarro's movements and adopted an attitude of ambivalence toward the small band of curious bearded men.

On November 15, Atahuallpa learned of the arrival of the Spanish in nearby Cajamarca and announced that he, too, would enter the city the next day at sunset to deal with the invaders personally.

About noon of the following day, 400 uniformed attendants entered the main plaza and began to clear away all uncleanliness in preparation for the royal visit. All the while Pizarro and his men watched in hiding and awaited the arrival of the Inca chief.

At sundown an awesome assemblage of 5,000 Indians swept down on the city from over a nearby hill. In the very midst of a hundred hand-picked bodyguards, soldiers, relatives, and local chiefs was Atahuallpa in all his regal magnificence. He ordered the imperial litter moved to the center of the plaza and waited for the confrontation.

From inside one of the buildings a lone Spaniard emerged in the company of an interpreter. He was a priest by the name of Friar Vincente de Valverde and was the spokesman for Pizarro's group. He claimed that he had come to visit this land for the purpose of converting the natives to the true religion. Immediately after this, he instructed Atahuallpa to step down as king and pay homage to Charles V. Friar de Valverde then held up a bible and waved it in the Inca's face, insisting that there was no god but Jesus Christ. Outraged, Atahuallpa ordered that the book be brought up to the litter for his inspection. It appeared to him to be nothing but a collection of pressed coca leaves and at best a puny symbol of the priest's power.[8] Contemptuously he tossed the book to the ground and announced to the priest that he was there to reclaim his people's stolen property and deal with the invaders as he saw fit.

The Spanish priest picked up his bible and fumbled back to the building shouting, "Fall on. . . . I absolve you." Atahuallpa interpreted this as the act of a craven man and addressed the large gathering in the central courtyard, saying that the bearded men were cowards and that they had already capitulated.

At once there was a sound like thunder, and many men began to fall to the ground all around the Inca chief. From their hiding places in the buildings, Pizarro and his men had lashed out suddenly and ruthlessly. They fired cannons, then charged on horseback into the midst of the crowd. Other Spaniards on the rooftops fired their rifles at will into the hapless Incas, who had never before even heard the sound of a gun or seen a horse.

The battle lasted only twenty minutes and ended as one of the most astonishing victories in military history. One hundred and sixty-eight weary European soldiers and adventurers killed an estimated 2,800 Incas without losing a single man. Incredibly, Pizarro himself was the only member of his group wounded, having been

**Sixteenth-century representation of Inca coca abuse.** ***(Courtesy of Felipe Guamán Poma de Ayala)***

stabbed in the arm accidentally by one of his own men while in the act of kidnaping Atahuallpa. It has been suggested that the reason the Inca soldiers were so ineffective in resisting the Spanish is that they were all drugged with coca.[9] Sensational speculation of this kind is often a product of the ignorant or the misinformed, and circulation of historical second guesses such as this can amount to slander. It seems to imply that if coca is a drug plant, then its action must be just like that of opium, and goes on to assume that coca use would tend to thicken or confuse any situation, especially one that called for clear thinking and fast action. Actually, the mildly stimulating effects of the leaf may have proved to be exactly what the Inca soldiers needed to resist the shock of the ambush and encourage their own military efforts.

Pizarro used the captive Inca as an object for ransom, and did indeed collect a fabulous amount of gold and silver from the frantic population. After all requirements for the ransom were met, Pizarro went back on his word and turned his vengeance on the Inca chief.

In August of 1533, Atahuallpa was executed by Pizarro in the main square of Cajamarca and the Inca reign was suddenly and violently brought to an end. The great Inca empire, known to millions for the past four centuries as *Tahuantinsuyo* (the four corners), was divided and renamed.

The entire country immediately fell into a state of uncontrolled chaos. Word of Atahuallpa's death spread quickly and was received with a different reaction wherever it went. Traditional enemies of the Incas seized the occasion to declare war, while others grieved to the point of suicide. The news sparked dissatisfied peoples everywhere to revolt and declare their independence. There were mass migrations, looting of graveyards, neglected crops, and famine for the first time in centuries. The wild disorder that came with the Spanish ruined Peru and its people. Drunkenness became a common public disgrace and coca-leaf use experienced a vicious prohibitional backlash that caused habituation to the leaf to become widespread and excessive.

# 5

# Imperial Spain and the Inca Leaf

After the death of Atahuallpa, the combined forces of Francisco Pizarro and Almagro marched out of Cajamarca and pushed their conquest to the very heart of the Inca empire. The Spanish forces moved south and followed hundreds of miles of good Inca roads through Jauja and Vilcashuaman. All along the way they were told of Cuzco, the golden city to the south. For Francisco Pizarro, this could only be his Panamanian dream of El Dorado come true.

After a brief skirmish in the hills about the city, Pizarro and his men marched victoriously into the great Inca capital. Cuzco was later described to Charles V as "the greatest and finest city ever seen in this country or any where in the Indies. . . . It is so beautiful and has such fine buildings that it would be remarkable even in Spain."[1] The Spanish conquistadors entered the city with little resistance and quartered themselves in the main square, waiting for the counterattack that never came. Pizarro took up residence at the Casana, the former palace of Pachacuti, and set about the business of planning the rape of Cuzco.

One of the early targets of the Spanish soldiers was Coricancha, Pachacuti's sacred enclosure in the temple of the sun. Inside they discovered the old king's garden with its delicate golden replicas of maize and coca plants. These items and other irreplaceable art treasures were taken and melted down into crude bullion before the year was out.

A member of the Spanish party named Sancho described vast stores of coca in Cuzco that were apparently of little interest to the Europeans, for they allowed these huge warehouses to be plundered at will by the natives during the general license that followed them into the city. Later it is known that some of the Spanish conquistadors began chewing coca secretly, depending on the Indians for their leaves.[2]

When Pizarro and his men entered Cuzco in November of 1533, the Spanish colonial rule that they represented had already been alive in the New World for over a generation. From their previous experience with American Indians farther north, the Spanish realized that they must successfully colonize if they wished to make their land acquisitions permanent. They therefore encouraged long-term settlement by awarding Spanish citizens in the New World large land parcels that came equipped with groups of natives who were forced to work the land for them. These human allotments were termed *repartimientos* and were the property, or *encomienda,* of the European settlers.

Twenty-five years before Pizarro reached Cuzco, Charles V outlawed the *encomienda* system on the basis of morality, only to change his mind a short time later at the request of Hernando Cortez, whose conquistadors were having difficulty subjugating the Aztecs in Mexico. In 1526 the *encomienda* was reestablished in Mexico and repeated in Peru three years later. As a direct result of this imperial edict, thousands of Incas were sent to the mines as slaves to satisfy

**Mama Coca gives the leaf to the Old World.** ***(Courtesy of the Fitz Hugh Ludow Memorial Library)***

the Spanish desire for precious metals. As the discoveries of new gold and silver mines increased, the need for Indian laborers to work in them grew proportionately until it became clear that the Spanish would have to resort to something even more devious than the *encomienda* to coerce enough natives to descend the narrow mine shafts.

They experimented with several methods but finally seized upon the Indians' love of coca and the laws of supply and demand. First, the Spanish coca monopoly made the leaf so expensive that practically no one could afford it; then they announced that they would issue free coca as a ration to those who agreed to work in the mines.

In spite of the fact that a great deal of the king of Spain's fortune was being chipped out of the mountains by coca-chewing Indians, his government actively denounced the natives and joined with the Church in a war against the use of coca.

The Catholic Church was a very powerful force in the early conquest of Peru, and was especially relentless in its pursuit of coca. It viewed the use of the leaf as an idle and unnecessary luxury and voiced numerous objections to its continued use. It was officially condemned by the Ecclesiastical Council of Lima in 1551 and again in 1569.[3] There was even an unsuccessful attempt by Church bishops to exterminate the plant completely by pulling them all out at the roots. These gross overreactions suggest that the Church was painfully aware of the bond that coca formed among the natives and that its use in Indian ceremonies recalled a mood of national and religious unity. For this reason, the Spanish realized that coca constituted a sizable obstacle to the spread of Christianity and Spanish imperialism.

The Indians answered the charges of the priests in typical fashion by claiming that coca only sustained them in their work and actually benefited all concerned. The priests then argued that if coca sustained the Indian, it was of course a food, and as such, its use could not be allowed before taking the Holy Eucharist. Since coca was used (as a food item) during practically every waking hour, the Church reasoned that the mandatory precommunion fast could not possibly be honored and that this practice constituted a threat to the spiritual well-being of the Indians and should be forever disallowed.

After much debate on this subject, the Church decided that coca was not a food after all and so could not sustain; hence, the Indians' belief in the powers of coca was false. The bishop of Lima termed it *un delusio del demonio,* a delusion of the devil, and pressed the state to outlaw the plant on these grounds.[4]

The Spanish colonial government, on the other hand, was obliged to view the leaf in a somewhat different light. Pizarro's institution of the *encomienda* system not only awarded land parcels with mining rights to Spanish settlers, it also gave them title to the large Inca *cocals* on the eastern slopes of the Andes. These new land barons immediately recognized the local demand for the leaf and reacted to it by almost doubling the number of plantations and increasing production in them to all-time high. In 1569, during the reign of the fifth colonial viceroy, it was reported that over 2,000 Spaniards had interests in the coca trade and that the leaves they produced were purchased by more than 300,000 Indians.[5]

As might be expected, the Spanish merchants soon realized an enormous profit from the leaf on the basis of its popularity and were well aware of the fact that it had grown to represent a sizable part of the Peruvian economy.

Pedro Cieza de Leon, who compiled a much-respected journal of events in the years after conquest, noted that "in the years 1548, 1549,

**A coca caravan as seen by Spanish conquistadors.** *(Courtesy of Theodore de Bry)*

1550 and 1551, there was not a root nor anything gathered from a tree . . . which was in such great estimation. In those years they valued the *repartimientos* of Cuzco, La Paz and Plata at $80,000 . . . all arising from coca." He went on to add, "There are some persons in Spain who are rich from the produce of this coca, having traded with it . . . sold and resold it in the markets."[6]

With this amount of money involved in the coca trade, a powerful Spanish lobby surfaced to support and defend the leaf. They insisted that its use was too widespread to suppress and pointed to its economic role in the new system. They claimed that the Spanish-owned coca alone could inspire the Indians to participate in the system by forcing them to purchase the leaf with cash; instead, the Indians made coca their money and used it in place of cash to obtain other goods and services. The pro-coca lobby deliberated with the Church for years, constantly debating the motives for using the leaf. The wrangle at last settled into a comfortable if not delicate compromise whereby the Church agreed to ease its attack on coca, officially persuaded that the Indians were prompted to use the leaf out of necessity rather than vicious gluttony or devil worship.

Paradoxically, the Indians later paid the Catholic priests coca in lieu of money for indulgences, and were forced to supply it as a sacrificial agent for Catholic feast days and Sundays. The relationship between coca and the clergy became so extensive that eventually a tax of one-tenth of the entire coca crop was set aside for the Church; the situation later degenerated to the point where the greater part of the revenue of the bishops of the cathedral in Cuzco was derived from coca.[7]

Although the decision to allow the Indians to

**A Spanish *cocal* in the *montaña*. *(Courtesy of A. A. Moll)***

continue to use the leaf could be entered as an example of the Spanish doing the right thing for the wrong reason, there is a strong element of truth in their claim that the Indians actually needed to chew coca. After the breakdown of the Inca system, the food-exchange cycle that kept the Indians nourished fell apart, and many Incas must have been forced to turn to coca as a remedy for hunger pains. It had long been used as a famine food in times of drought, crop failure, or earthquake. This time a great man-made disaster forced the people to use coca as a surrogate in the absence of their regular diet.

Those newly conquered highlanders assigned to work in the *cocals* soon learned why Inca criminals were sent there as punishment, for it is estimated that between a third and a half of all the Indians assigned to work in the *cocals* died as a result of their five-month service.[8] The climate in the lower valleys was disastrous for the alpine Incas, whose lungs had been enlarged by evolution to breathe thin air. They were almost completely helpless in resisting the *verruga,* a fatal disease characterized by eruptive nodules and severe anemia. *Mal de los Andes,* also called the *uta,* claimed many lives; it was a cancer-like disease that destroyed the nose, lips, and throat. By this time, the native population had dwindled to two million or less from an estimated peak of ten million in 1500![9]

Finally, on October 3, 1572, Viceroy Toledo published a series of ordinances designed to govern the conduct of the plantation owners and protect the Indians. Some of the corrective measures to be enacted as a result of these documents point to the fact that Spanish abuse of the

Indian population was widespread and in great need of some legislative attention.

The decree stated, in part, that no Indian was to be forced to work in the coca plantations against his will, nor was he to be required to pay any tribute to the landlord in coca leaves. Those who overburdened Indians on the way to the *cocals* were to be fined heavily. Since His Majesty had ordered that the Indians were not to be used as beasts of burden, owners were to be sure to apply this ruling to the coca traffic as well. All roads and bridges in and out of the *montaña* were to be repaired and maintained by the owners of the coca plantations, and they were to see that there were shelter huts erected along the sections of roads crossing the mountains. Since many of the valleys suitable for coca belonged to other tribes, such as the Camayos, it was now forbidden to take these valleys from them or to work them against their will; these Indians were also not required to bring wood, eggs, or fowl to the plantation owners in order to get work. Older Indians were not to be assigned excessive tasks, but were to be allowed to work as they could. The owners were to see that the Indians were given food on the days when it rained and during the time when the leaves were ripening. The Indians were only to work for twenty-four days at a time and then not before sunrise, after sunset, or on Sundays or holy days. The official ordinance also stated that no Indian woman was to be forced to labor on a plantation against her will, and in case she wanted to work, she was to receive pay.[10]

If the slave labor in the *cocals* was bad, the condition of the mine workers was worse. Writing to the Council of the Indies in 1550, Domingo de Santo Tomas said, "Some four years ago, to the complete perdition of this land, there was discovered a mouth of hell, into which a great mass of people enter every year and are sacrificed by the greed of the Spanish to their 'god.' This is your silver mine called Potosi. So that your highness can appreciate that it is certainly a mouth of hell . . . I will depict it here. It is a mountain in an extremely cold wasteland, in whose district for a distance of twenty-five miles, no grass grows, even for cattle to eat, and there is no wood to burn. They have to fetch food on the back of Indians or the llamas they possess. . . . The wretched Indians are sent to this mountain from every *repartimiento*. . . . No one who knows the meaning of liberty can fail to see how this violates reason and the laws of freedom. For to be thrown by force into the mines is the condition of slaves or of men condemned to severe punishment for grave crimes."[11]

Because the need for coca was greatest where the greatest amount of work was done, the price charged the Indians at the mining towns was significantly higher than anywhere else. Acosta reported that the coca trade at Potosi was worth the equivalent of $500,000 per annum and that in 1583 the Indians consumed two and one-half million pounds of leaves that were roughly twice as expensive at Potosi as at Cuzco.[12] The Spanish soon realized that at both the mining towns and the plantations, work could not continue without the help of coca. When the price became too great for the Indians to bear, the owners were forced to provide the leaf for them just as fuel might be given to an engine to make it do a desired amount of work.

The deplorable exploitation of coca by the Spanish directly inspired at least two Indian revolts, one in Quito and the other near La Plata, which later became known as the uprising of the Catari brothers. The three brothers led thousands of outraged peasants against Joaquin Alos, an unscrupulous local bureaucrat who had a history of economic abuses against the people. Alos's constant lies and dirty tricks to gain personal wealth were well known in the district even before his appointment as *corregidor* in 1778. It is only after he tried to establish an outrageous financial monopoly over the sale of coca that the Indians threatened to cut off his head and drink *chicha* from his skull. Both sides then escalated the situation into a small war that ended with the capture of Nicholas Catari and Simon Castillo by the Spanish. They were charged with being the principal instigators of the uprising and in an effort to crush the revolt, the two were sentenced to public execution on May 7, 1781. When Catari and Castillo reached the hangman's gallows, they stopped, filled their mouths with coca, and chose to die chewing the leaf.[13]

The Viceroy Marquis of Canete finally acknowledged the abuse suffered by the Indians at the hands of greedy landlords and recognized the almost intolerable hardships endured by the workers in the *cocals*. As a result, he was inspired to issue even more royal decrees that once again prohibited forced labor in the coca plantations and insisted that all *corregidores* work to protect the health and security of the coca workers. Making a move most probably designed to placate the Church, he tacked on an amendment outlawing the use of coca in witchcraft. Unhappily, Canete's laws concerning coca, like those before them, were all but ignored and the situation continued to get worse.

The effect of Spanish control over coca was cataclysmic. The leaf that had served the Indians in one way or the other for over 4,000 years was now only serving to submerge them deeper into a strange, new feudal system and hasten the end of their way of life.

The Incas were the natural inheritors of the coca leaf. It was their ancestors who first discovered the plant, experimented with it, and worked through its acceptance and application in the society. They were the consummation of the Andean Indians' kinship with coca, a most refined version of man's cultural symbiosis with a plant.

At the same moment that the Spanish wrenched the coca plant from the Peruvians, they delivered it unwanted into the hands of unprepared Europeans, who received it in ignorançe. In the Old World, the leaf would have to grow up all over again, slowly being rediscovered by a variety of interested individuals and incorporated into their culture during the seventeenth and eighteenth centuries.

# 6

# Pirates, Priests, and Poets: Coca in Western Europe

In 1533, Hernando Pizarro set sail for Spain with the first cargo of treasures from Inca Peru. He was eager to display these riches and curios to the court of Charles V in order to gain support for the new conquests in the Americas.

The initial transfer of goods from South America to the European continent contained the "royal fifth" of Atahuallpa's ransom money, examples of weapons, garments, foods and plants, and included specimens of coca.[1]

While the Spanish aristocracy could not help but be curious about a leaf that could supposedly conquer hunger and fatigue, they were, at the same time, suspicious of any Indian cures and hesitant about accepting the stories of coca's power at face value. They were the products of a type of thinking that had five years earlier condemned the potato plant because it was not mentioned in the Bible. Ladies and gentlemen of the court must have found the practice of holding twigs or leaves in the mouth alien to their post-Renaissance attitudes of etiquette.

Imperial Spain was in its finest hour. Charles V of Hapsburg had recently added the title and power of Holy Roman Emperor to his vast inheritance, which at this time included land in Austria and France, most of Italy, all of Portugal, two huge American colonies and, of course, the Spanish crown.

His aristocratic circles in Spain were pretenders at religious devotion and were driven in their efforts to imitate or surpass the Italian models of courtly formality and sophistication. The coca leaf made its debut in Spanish Europe during this time of strict social and religious conformity; with the "Holy Inquisition" just ahead, it is understandable that no real interest was generated in a South American Indian chewing custom that had significant indigenous religious implications.

By this time, the Inca religion had "officially" been replaced by Roman Catholicism, but continued to flourish despite the objections of the Spanish priests.[2]

The people of Peru who used the leaf, like their predecessors before them, lived in a complicated world full of magic spirits and superstitions. It was well known, for example, that Atahuallpa took great care to destroy virtually everything he touched to avoid having these objects manipulated by sorcerers. Peruvians arranged their existence according to prophecies made by men who studied the entrails of animals and clouds. The auspicious rituals that took place throughout their daily lives must have appeared peculiar if not blasphemous to most Europeans. Because coca was such a key ingredient in these numerous events, it follows that the Spanish associated it with the unusual and tended to ignore its worth.

At roughly the same time coca was introduced to the court of Charles V, there was also a proliferation of the leaf as a trade item to Spain's other holdings in the New World. Unlike the royal scrutinization coca suffered in Europe, some people in Central America and Mexico seemed to have welcomed the leaf and easily incorporated it into their ways. The reference to coca in Francisco Hernandez's work on the materia medica of Mexico is a case in point. Hernandez studied medicine in Spain and practiced at the famous monastery of Guadalupe in Extremadura. In 1567, Phillip II made him his personal physician and three years later appointed him Protophysician of the American possessions. Hernandez describes drug use in Mexico at this time, and his writings are thought accurately to represent Mexican beliefs about their properties. Mentioning coca, Hernandez says, "It extinguishes the thirst, nourishes extraordinarily the body, calms down hunger when there is not abundance of food and drink, and takes away the fatigue in long journeys. They use also [the leaves] mixed with *yetl* [tobacco] as a pleasure, when they remain in their villages to induce sleep, or obtain intoxication, peace and oblivion of all their sorrows and cares."[3] Hernandez does not specify whether the Mexicans chewed the mixture or smoked it.

The writings of Father Bernabe Cobo, S.J., suggests that despite the official denunciation of coca by the Church and state, there were at least a few Europeans who were not afraid to make practical use of this valuable drug plant. Father Cobo studied the history of the Incas for ten years in Cuzco and in that time undoubtedly became quite familiar with coca. When a fellow priest suggested that Cobo make use of the "Inca leaf," he did not hesitate and later composed what could be the first written testimonial to the anesthetic properties of coca. He said, "It happened to me that calling once a barber to pull me

**Hernando Pizarro presents the coca leaf to the court of Charles V.** ***(Courtesy of the artist, Darrell Orwig)***

*Of the Coca.*

I Was deſirous to ſee that hearbe ſo celebrated of the Indians, ſo many yeres paſt, whiche they doe call the *Coca*, whiche they doe ſow and till with muche care, and diligence, for becauſe they doe vſe it for their pleaſures, which we will ſpeake of. The *Coca* is an hearbe of the height of a yerd, little more or leſſe, he carrieth his Leaues like to *Arraihan*, ſomewhat greater, and in that Leafe there is marked an other Leafe of the like forme, with a line very thinne, they are ſofte, and of coulour a light greene, they carrie the ſeede in clusters, and it commeth to be ſo redde when it is ripe, as the ſeede of *Arraihan*, when it is ripe. And it is of the ſame greatneſſe, when the hearbe is ſeaſoned, that it is to be gathered, it is knowen in the ſeede, that it is ripe and of ſome rednes like to a blacke kiſhe coulour, and the hearbe beyng gathered, they put them into Canes and other thinges, that they may drie, that it maie be keapte and caried to other partes. For that they carrie them from ſome high Mountaines, to others, as Marchaundiſe to be ſoulde, and they barter and chaunge them: for Mantelles, and Cattell, and Salte, and other

**The coca entry from *Joyfull News out of the Newe Founde Worlde wherein is declared the Virtues of Hearbes, Treez, Oyales and Plants.* (Courtesy of John Frampton)**

out a molar because it moved and hurt me considerably, the barber told me it was a pity to take it out because it was sound and healthy; and being present a religious friend of mine he advised me to chew coca for a few days. I did it and the toothache went away, and the molar became firm and like the rest. The sap of the coca strengthens the stomach and helps digestion."[4]

More than fifty years after Hernando Pizarro brought the original shipment of coca from the Americas, Dr. Nicholas Mondares of Seville, who was possibly the earliest coca experimenter in Europe, published the first academic account of coca, called *Historia medicinal de los cosas que traen de los Indias occidentales que sirven al uso de Medicina* (1580).* The manuscript was copied into Latin by Charles L'Ecluse, who was then the director of the emperor's gardens in Vienna. In 1596, John Frampton "made the book into English" and entitled it *Joyfull News Out of the Newe Founde Worlde wherein is declared the Virtues of Hearbes, Treez, Oyales and Plants.*

By the beginning of the seventeenth century, then, Western Europeans had access to some published material concerning coca printed in at least three different languages. While these works were available from Austria to England, their esoteric or intellectual tone limited circulation and kept the material away from the casual reader. In *Paradise Lost,* John Milton mentions a New World plant *(Ficus indica)* and Atahuallpa, but makes no mention of the divine leaf of the Incas.[5]

On the other hand, the poet Abraham Cowley, who was a contemporary of Milton's, not only knew about coca but was moved to write a long poem about it. Given the particulars of his extraordinary life, it is not surprising to learn of Cowley's familiarity with coca or his admiration for it. He had lived in exile in Paris for two years and there, was exposed to an endless parade of travelers passing through the court of Lord Jermyn, by whom he was employed as a secretary. Cowley's mysterious and eccentric career led him through a number of different roles, including physician, royalist, metaphysical poet and playwright, early member of the Royal Society, and even a secret agent for the Stewart descent line. He was also known to have consorted with the most famous "wits and gallants" of his day, suggesting that it was perhaps among this group that coca won its first battle for acceptance in Europe.

At the Restoration, Cowley moved from London to a retirement farm in the country where he could devote more time to his first love, botany. From his country home he produced *The Book of Plants,* and in it settled once and for all an argument between Bacchus and Ceres concerning the world's greatest plant:

> Of all the plants that any soil does bear, This Tree in Fruits the richest does appear, It bears the best, and bears them all Year. Behold how thick with leaves it is beset; Each Leaf a Fruit and such substantial Fare, No fruit beside to rival it will dare. . . .
>
> Our Virachocha first this Coca sent Endow'd with leaves of wond'rous Nourishment, Whose Juice Succ'd in, and the Stomach Tak'n Long Hunger and long Labor can sustain; From which our faint and weary bodies find More Succor, more they cheer the drooping mind Than can your Bacchus and your Ceres join'd. Three leaves supply for six days march afford; The Quitoita with this provision stor'd Can pass the vast and cloudy Andes o'er. The

*While Mondares is to be credited as the first academic publisher of coca information, he may also be responsible for fanning the flames of the coca debate when he declared that chewing coca caused the Indians to "go out of their wittes [sic]."

**Abraham Cowley, who consorted with the "wits and gallants" of his day, wrote a long poem about coca in the late 1600s.** ***(Courtesy of the Trinity College Library)***

> dreadful Andes plac'd twixt Witner's Store of Winds, Rains, Snow and that more humble Earth That gave the small but valiant coca birth. . . .
>
> Nor coca only useful art at Home, A famous Merchandise thou art become A Thousand Paci and Vicungni groan Yearly beneath thy Loads, and for thy sake alone The spacious Worlds to us by Commerce Known.[6]

Although Cowley calls coca a "famous merchandise," that may have been all it was, and then only among a select few. It would be another 150 years before Europeans would realize the value of coca and begin to use it.

Even after the publication of Cowley's poem, there was no immediate public attention focused on the leaf. For those who were interested, there was still a great deal of uncertainty surrounding its effects and confusion regarding its acceptance in European society.

Conflicting accounts of the use and abuse of the coca leaf were contained in the works of Casas, Acosta, Garcilasco, and Zarte on the one hand, and the Church and state on the other. Those reports that confirmed the powers of coca were generally considered fabulous and not taken seriously. The first purely botanical classification of coca, done by a man by the name of Leonardi Plukenetii, was not available until 1692.[7]

Apart from this select handful of aristocrats, scholars, poets, and priests, there was still another group of Europeans whose colorful history was mixed up with the coca plant at a relatively early date.

At the close of the sixteenth century, the Spanish monopoly of trade in the New World gave rise to a certain amount of envy on the part of other interested nations. As a result, pirates from France, England, the Netherlands, and even Mohammadan raiders from as far away as Algiers began arriving in the New World for the purpose of freebooting the cargo-laden Spanish galleons leaving Peru.

Spain's treasures from her New World colonies were taken from her on land as well as at sea. Peruvian gold, silver, and other trade items from western South America were brought to Panama by ship and then portaged across the isthmus to the Caribbean seaboard cities of Nombre de Dios and Puerto Bello, and from there to Seville. The road across the isthmus was called the "gold road" and was worked extensively by land-based buccaneers. Sir Francis Drake even landed his ship on Peruvian soil as early as 1579 for the purpose of helping himself to the new colony's exports.

Around 1650, Tortuga, one of the Spanish islands in the Caribbean, began to attract a large assortment of unconventional citizens. The island was soon full of Englishmen, Frenchmen, Africans, Portuguese, and Moors. There were escapees from the Jamaican sugar plantations, runaway slaves, and dissatisfied logwood cutters from the Bay of Campeachy in Mexico. Their collective presence amounted to a small army of men with little to lose; they soon assembled themselves into a sizable pirate fleet and made their living almost exclusively by raiding Spanish shipping.

Although their main targets were gold and silver, it is certain that these pirates did not hesitate to help themselves to other useful goods carried aboard the Spanish ships. By the early 1600s, quantities of coca leaves began to appear on the manifests of cargo ships departing Lima.[8] Many of these shipments fell into the hands of pirates who may have either used the leaves themselves or facilitated their distribution throughout the Spanish Main, or both.

The social behavior of these corsairs was not in any way hampered by the conventional European ideas of good taste or the religious and moral precepts of the Catholic Church. As a consequence, they may have been much more receptive to actually making use of the coca leaf, as they later demonstrated with their acceptance and early use of the tobacco leaf.

Their knowledge of coca's reputation for power and endurance almost could not have been avoided. Seventy-five years earlier seamen with Peter Martyr quoted Dominican monks who told of coca chewing and cultivation all along the Pearl Coast of Venezuela.[9] Christopher Columbus also noted coca use among the principal men of Panama.[10] In addition, he reported that coca use was prevalent among the Nicarao on the Pacific coast of Nicaragua.[11]

The pirate John Esquemelling wrote in his journal on Wednesday, January 26, 1680, "The Indians of this land [island of Iquique] eat much and often a sort of leaf that tastes like our own bay leaves in England. . . . The leaves of which we make mention above are brought to the island in whole bales and are given to each man a short allowance."[12]

**A Caribbean pirate.** ***(Courtesy of Howard Pyle)***

It appears that the practice of chewing coca leaves for one purpose or another had spread quickly throughout a great portion of New Spain by the late 1600s, and by this time at least one account even suggests that pirate ships moved the leaf as far away as the North American settlements.[13] Its availability to Indians both in and out of the Inca world was now almost entirely in the hands of Spanish merchants and pirates who both recognized its value as a trade item and treated it accordingly. As early as 1650, Ecuadorian Indians were forced to abandon their coca customs completely, not because the leaf was unavailable to them, but only because by this time the white man had made it too expensive.[14]

While coca use seemed to be on the increase in America, the Europeans remained largely uninterested and the demand for the leaf in the Old World was confined to a few. Access to shipments of coca also presented a problem to potentially interested parties, as the Spanish still ran a tight monopoly of Peruvian trade items.

At this time, Western Europe was filled with wizards, mountebanks, and other medical charlatans all claiming to have the secret of a long life and powers over pain. Van Loon describes one man wrapped in a long red cape claiming to be a prince of Babylon who had discovered King Solomon's "elixir vitae" among the ruins of Jerusalem.

Those Westerners who decried coca use as an example of pagan gullibility were the same people who had developed a long tradition of esoteric curing rituals and substances of their own. At this time, the University of Cracow in southern Poland was offering a degree in Black Magic, and some degree holders traveled throughout the European countryside administering so-called healing potions.[15] These potions were administered from their medicinal pouches and one physician's bag from this period contained horse teeth, part of a weasel's skeleton, a lynx's hide, a serpent's tail, a twig of mountain ash, a bird's windpipe, and a few round pebbles.[16] Popular pain remedies of the day included menstrual blood, the moss scraped from the skull of a man killed by violence, powdered mummies, and crushed jewels. Coupled with such remedies was the widespread practice of selling relics of dead saints to insure a painless death and entrance into heaven. European pain-killing practioners even resorted to a literal translation of the last chapter of St. Mark to save them from worldly suffering; the phrase, "They shall lay hands on them and they shall recover" truly took on extra scriptural meaning for many Europeans when feudal kings began touching their subjects in an attempt to cure them.

Henry of Navarre left his carriage repeatedly to make the sign of the cross over his people's injuries, and it is a fact that Louis XIV touched nearly 2,500 Parisians as a medical prophylactic against the plague.

One of the principal reasons that Europeans entertained such trickery was that medical science had advanced to the point where surgery had become a common and fairly successful practice that had to be performed in the absence of any kind of anesthesia. This situation effectively turned the operating theater into a torture chamber and sent millions of sufferers on a desperate search for any way to end the agony of surgery and healing.

It is both interesting and ironic to learn that in the midst of all this suffering, special monks knew the secret of administering sleep sponges to patients in pain and Paracelsus used laudanum (an opium drink) to bring on deep, relieving sleep. Father Blas Valera had also made public the first European medical application of coca in 1609[17] when he said, "Coca preserves the body extraordinarily from many infirmities and our doctors use it powdered for application to sores and broken bones, to remove cold from the body or to prevent it from entering, as well as to cure sores that are full of maggots. It is so beneficial and has such singular virtues in the cure of outward sores, it will surely have even more virtue and efficacy in the entrails of those who eat it."[18]

In spite of these good examples, narcotics from the plant world were almost universally considered untrustworthy as remedies and were actively avoided by most physicians.

In the mid-1600s, a barber-surgeon named Bailly of Troyes was fined heavily for stupefying patients with herbal potions and the practice was afterwards forbidden in France. In 1638, cinchona, the malaria-preventing bark, was available to anyone in Europe who cared to use it, yet the "Jesuits' powder" suffered the worst kind of denunciation at the hands of the most enlightened physicians of the day, and was ignored. Oliver Cromwell was offered the powder on his deathbed, refused it, and promptly died the next day.

Difficulties also came to bear in the Calvinistic belief that God sent pain to atone for sins and any interference by man could only involve even more sinning and consequently more pain! Many doctors simply resigned themselves to the fact that pain and surgery were inseparable and all a part of God's plan for existence.

Although there were accounts of patients helped by plant remedies, there was an equal number killed by overdoses of one kind or another. The reason for this is that the anesthetic dose is often very close to the lethal one. For the sake of uniformity of potency, there developed a tremendous need to discover the exact property of the plant that gave it its power, and a considerable amount of effort was put forth to locate and even separate the working parts of *Cannabis, Papaver somniferum,* and mandrake. Unfortunately this job fell into the hands of the practicing alchemists of the day, whose only credentials were their wild experiments with the transmutation of metals and their efforts to separate gold from mercury and other nonrelated substances. This absurd mixing of necessity with magic and economic adventure through chemistry produced an equally absurd theory that some yet-to-be-discovered element in gold would surface and put an end to human suffering.

During the Middle Ages, man had turned to every source he could think of in an effort to spare him the agony of physical pain. Religion, royalty, and alchemy seem to have been the most popular. In the more extreme cases, patients asked to be knocked unconscious or volunteered to consume a near-fatal dose of alcohol.

Curiously, each time a genuine anesthetic was introduced in Europe (and all the major ones

**A coca scholar of the Middle Ages.** ***(Courtesy of the Fitz Hugh Ludlow Memorial Library)***

were at a relatively early date), they invariably suffered the stamp of severe social disapproval.

Raymond Lully, who discovered sulphuric ether as early as 1290, was stoned to death for his efforts in Tunis in 1314, and his anesthetic discovery apparently went with him. Two hundred years later, Philippus Aureolus Theophrastus Bombastus von Holenhiem, also known as Paracelsus, gave Europe another great painkiller when he made laudanum, or tincture of opium, by softening dried poppy juice with boiling water and spirits of wine. His laudanum contained one percent morphine, which is still the world's leading painkiller. Paracelsus was the object of fierce criticism throughout his whole life and was allegedly fired from Basel University because he "employed newfangled medicine . . . and actually cured patients."[19] A short time after this, the city council of Basel ran him out of town. Paracelsus was killed in a drunken altercation in Salzburg in 1541 and his death marked the end of meaningful experimentation with laudanum. Joseph Priestley, who correctly applied nitrous oxide toward the relief of pain in the early 1700s, had his house burned to the ground by an angry mob.

The other major anesthetic agent "asleep" in Europe at this time was coca. Like the others, it was overlooked despite early medical announcements that suggested success in application. These accounts, coupled with testimonials such as Father Cobo's, should have been enough to inspire serious interest in the anesthetic properties of the leaf, but they did not. Instead, Cowley's "famous merchandise" was first snubbed by courtiers of Spain, whose airs of pious recitude could not allow space in their world for an Indian herb, and then was almost completely ignored for almost an additional 150 years while people in pain searched everywhere for relief.

# 7

# The Age of Enlightenment and the Isolation of a Fugitive Alkaloid

At the turn of the eighteenth century, major changes began to unfold that would revolutionize man's association with drug plants and especially coca.

Unquestionably, the greatest of these changes took place in the collective consciousness of the European people. The time now referred to as the Enlightenment produced a mood that called for a new approach to old problems, and men such as Newton, Descartes, Linnaeus, and Bacon appeared to sustain this new approach. Together, they established an atmosphere that was alive with new and powerful ideas. They created a science that began to make alchemy, philosophical guessing, and the miracle obsolete.

The sudden popularity of science was enormous; tabloids and magazines glorified the heroes of science and dilettante societies appeared all over Europe with plans for elaborate explorations in the name of the new order. Special fields of studies were being created to coordinate the incoming data and it was not long before European intellectuals began to realize the chemistry, biology, and medicine were all very closely related.

Herman Boerhaave, a Dutch physician, was the Enlightenment's gift to coca. In addition to being the foremost medical practitioner of the 1700s, he shares with Lavoisier the title of 'father of organic chemistry.' Boerhaave overturned Galen's doctrine of the four humors and was instrumental in rescuing modern medicine from the influence of the ancient texts. His teaching attracted pupils from all over Europe and his scientific researches were considered so important that they were translated into practically all the major languages of the day and read all over the world.

In 1708, Boerhaave published *Institutiones Medicae,* which contained a review of some herbs thought to have medicinal value. In it, he lists coca as one of the principals and discusses the plant in a way that would prove revolutionary both to coca and all science. Boerhaave set the course for some of the most important research in the history of medicine when he referred to the bitter constituent in the juices produced in the chewing of coca. He spoke of the leaves as housing a vital strength and suggested that there was a source of authentic nutritive value. By making these remarks about coca, Boerhaave gave science its first hint concerning the existence of alkaloids and launched an elaborate search for the working parts of active plants. Ironically, when most of the major plant alkaloids were eventually separated in the nineteenth century, coca, the plant that began the search, was the last to have its alkaloids identified and separated.

This same subject had completely confounded

**Herman Boerhaave.** *(Courtesy of the Fitz Hugh Ludlow Memorial Library)*

the alchemists. Their efforts to unlock the power of plants were disastrous and had to be dismissed as another spectacular failure. The entire notion was in danger of fading away completely until a man of Boerhaave's reputation and learning surfaced to reintroduce the subject and phrase it in a proper scientific context.

At the same time that he began modern science's search for alkaloids, Boerhaave also initiated an important shift in the public's attitude toward coca. His explicit references to the leaf's effect on humans was regarded by some as official assurance that the coca plant really did possess some remarkable properties. Boerhaave's scientific affidavit allowed people the luxury and excitement of contemplating the hidden mysteries of a strange new herbal that had associations with distant sun kingdoms, and they felt free to puzzle over its unanswered scientific questions. The Inca leaf was now beginning to be a real entity for the curious, science-minded citizens of eighteenth-century Europe, and it was only natural that it became the subject of many sensational reports sent back from New World explorers and scientists.

The eighteenth century not only generated a scientific effervescence that stimulated interest in coca, but also produced a unique political arrangement that allowed scientists from all over the world to visit Peru and study the leaf firsthand. Since Pizarro's conquest, it had been next to impossible for anyone to settle in, visit, or even trade with Spain's possessions in the Americas.

This situation began to reverse itself in 1700 when Phillip of Anjou, the grandson of Louis XIV, obtained the Spanish crown and put an end to the trade monopoly that existed on American goods. The new commercial policies of the Bourbon king granted both France and Great Britain trade privileges and the option to establish American colonies of their own. This detente in turn opened an era of brisk trade between all of continental Europe and Peru, and effectively put an end to piracy in the Caribbean.

Spain's territorial losses in America were small. She gave up only unoccupied islands at first and later, among the larger islands, only Jamaica to the English and Western Hispanola to the French.[1]

Both France and England turned their newly acquired lands into sugar-cane or tobacco plantations and cultivated them with black slaves from Africa. The Europeans soon discovered that native Caribbeans were simply not a physical match for the hardships of forced plantation work in the tropics, so they were replaced by the captive Africans. Because of this policy, the Carrib Indians eventually vanished from their traditional homelands altogether.

The English, who actually introduced themselves into the area as slave traders, had established a receptive market for slaves all along the Atlantic seaboard of North America, as well as the Caribbean, dating back to 1502. There are reports that some plantations serviced by English slavers were also supplied with coca.[2] The design, of course, was to encourage the African slaves to chew coca and presumably increase production following the Spanish model with the Incas in Peru.

One of the most infamous of these slave plantations was established on the island of Jamaica. In a very brief period, the English pushed aside the native population and filled the island with impressed laborers from Guinea. Conditions were notably bad and the work uncommonly hard; given these facts, it is not surprising to find coca listed among the earlier plants introduced to the island in a book published in 1756.[3]

Under King Phillip's new trade legislation, ships of many nations were able to anchor in the Peruvian harbors, take on loads of coca leaves, and market them wherever they pleased.

In 1712, only four years after the laws were relaxed, at least nine French ships visited and traded in Chile and Peru.[4] The variety of new and exciting goods and information brought back to Paris from America inspired the French government to sponsor a large scientific expedition to return to Quito in 1735.[5] The principal investigator was the mathematician La Condamine, a friend of Voltaire's, who wanted to measure an arc of the meridian in the vicinity of the equator and thereby verify the shape of the earth.

The French government seized this opportunity to send other scientists along on the experiment and Joseph de Jussieu, a biologist and member of a large family of distinguished plant scientists, was among those chosen to make the trip. While only one member of the de Jussieu family accompanied La Condamine to South America, at least three of them collaborated in classifying the coca plant.

When he reached Quito in the spring of 1735, Joseph de Jussieu immediately set off on foot into some of the most rugged, unexplored territory in the world. He started southeast from Quito and did not stop until he reached the forests of Santa Cruz de la Sierra, almost 2,000 miles away.

In 1749 he was almost killed attempting to reach the famous coca plantations in the *yungas* of Corico in Bolivia. He managed to survive, however, and from this region he shipped samples of the coca plant to his brother Antoine in Paris. Only a year later, de Jussieu decided to return to Paris to complete his work and packed fifteen years' worth of botanical gatherings into huge crates for the trip back to France. Unhappily, de Jussieu's life's work was stolen off the loading docks and later dumped into the ocean by dissatisfied thieves.

The shock of this event eventually ruined de Jussieu, but at the time it did not prevent him from trying to carry on; he attempted to start all over again and elected to remain in the New World indefinitely to try to do so. Finally, twenty years later, he had to be taken back to France because he had gone mad and was no longer in control of himself. Needless to say, de Jussieu was never able to complete his work and he died a few years after his return, leaving numerous unfinished manuscripts.

Although the majority of de Jussieu's choice specimens were scuttled into obscurity, the coca plant he sent to his brother Antoine only a year before the tragedy has survived as probably the most famous single coca shrub ever gathered. Antoine continued his brother's work by accurately describing the leaves, placing them in the Malpighiaceae family of the genus *Sethia* and, most importantly, preserving them in the Museum of Natural History in Paris. The noted botanists of the day, including de Candolle and Linnaeus, used these same leaves for all their descriptions, mainly because they were the only ones available in Europe at the time.

After examination of the de Jussieu leaves, Carl von Linnaeus placed coca in the genus *Erythroxylon* of the *Erythroxylaceae* family; he modified Antoine de Jussieu's classification because of certain characteristics of the flowers. Linnaeus's names won popular favor and inspired a third member of the de Jussieu family (Antoine Laurent), the nephew of the two brothers officially to reclassify the species as *Erythroxylon coca*. Lamarck accepted the Antoine Laurent de Jussieu classification and copied the name in his classic *Encyclopédie Méthodique Botanique*. Although Lamarck's name appears after *E. coca* in many academic texts, it was the combined efforts of three members of the de Jussieu family that produced the accepted botanical binomial nomenclature for the coca plant.

Also included in the La Condamine expedition were two young Spanish mathematicians, Jorge Juan y Santacilia, age twenty-two, and Antonio de Ulloa, age nineteen. They were sent at the request of the Spanish king and distinguished themselves by making careful observations of depths, currents, and geographical features along the coastline. Their greatest accomplishment, however, was the compilation of many journal volumes filled with fascinating ethnographic accounts.

Juan and Ulloa reached the great trading center of Puerto Bello, Panama, in the summer of 1735 and were amazed to find it almost completely populated by blacks and mulattoes; they managed to count only thirty white citizens.[6] They commented on the city's reputation of being one of the unhealthiest ports in the Spanish

Main and were later informed that they would be lucky not to lose many men to disease, even after a brief stay.

The heat was oppressive and the mosquitoes swarmed in uncontrollable clouds everywhere. Local residents suffered from *peste,* a disease characterized by delirium and vomiting blood, and *bicho,* probably gangrene of the rectum, for which the prescribed cure was the insertion of a suppository of gunpowder, guinea pepper, and a peeled lemon.[7] They also noted an overabundance of women, and it was said that males began to deteriorate and die after thirty.

Understandably, the La Condamine expedition was anxious to leave Panama as soon as possible. On February 21, 1736, they set sail for Guayaquil aboard the frigate *San Cristobal,* reaching Quito approximately three months later.

**A coca plant drawn from the de Jussieu specimen.** ***(Courtesy of the Fitz Hugh Ludlow Memorial Library)***

During their extended visit in this city, the two young Spaniards recorded many details of everyday life among the local residents. On coca, Ulloa reports, "The use the Indians make of it is for chewing, mixing it with chalk or whitish earth called *mambi.* They put into their mouths a few cocal leaves and a suitable portion of *mambi,* and chewing these together, at first spit out the saliva which that mastication causes, but afterwards swallow it, and thus move it from one side of the mouth to the other until its substance be quite derived, then it is thrown away, but immediately replaced by fresh leaves. . . . This herb is so nutritious and invigorating that the Indians labor whole days without anything else, and on the want of it they find a decay in their strength." He also added that it preserved the teeth and fortified the stomach. After making these observations, Ulloa unaccountably confuses the issue by saying, "It is exactly like the betel of the East Indies. The plant, the leaf, its qualities and the manner of using it are all the same."[8]

The La Condamine expedition began as a geophysical trip to South America but ended as much more. Juan and Ulloa, together with the other members of the French party, dispersed and continued their scientific investigations throughout the continent. They revitalized interest in South American geography, anthropology, economics, and botany and can all be considered characters of paramount importance in stimulating the great scientific work of the nineteenth century.

Close to the turn of the century, Peru was visited by a man whom Charles Darwin called "the greatest scientific traveler who ever lived."[9] This was Friedrich Heinrich Alexander Baron von Humboldt, born in Berlin in 1769. Humboldt initiated the next important step toward a scientific understanding of the coca plant and, as so often happens in science, he achieved this honor by making a mistake.

By the time he was thirty, Humboldt was skilled as a linguist, biologist, and geologist. He was also an admirer of the La Condamine expedition and was anxious to take a firsthand professional view of the New World. His earliest opportunity to travel to America in the capacity of a scientist presented itself in the winter of 1799, when he learned of an expedition leaving France under the command of a Captain Baudin.

Humboldt, together with another botanist, Aimé Bonpland, applied for and received commissions with Baudin's group. Unfortunately, because of an obscure political dispute known to a few as the War of the First Coalition, France was forced to withdraw her financial support and, as a result, Baudin's expedition never set sail.

Somewhat discouraged, Humboldt and his companion decided to winter in Spain and it was there in the spring of the following year that they happened to meet Baron von Forell, the Saxon minister in Madrid, who was himself an amateur biologist. He told them that a friend of his, a high-ranking Spanish official, might make it possible for them to visit Spanish America at their own expense.

Events proceeded smoothly for Humboldt and Bonpland after that, and in a short time they were not only given permission to "execute all operations they should judge useful for the progress of the sciences," but were also authorized free use of Spanish instruments.[10] These unprecedented actions moved Humboldt to write, "Never had so extensive a permission been granted to any traveler . . . and never had any foreigner been honored with more confidence on the part of the Spanish."[11]

The interruption caused by the War of the First Coalition gave Humboldt an additional three months in Europe before he was scheduled to leave for the New World. He turned these extra days into an extensive period of preparation that included a visit with Hipolito Ruiz, who had conducted a detailed botanical survey of Peru just one year earlier. Ruiz even opened his collection to Humboldt and undoubtedly introduced the German scientist to the coca plant.

When any attention was focused on coca in the last days of the eighteenth century, controversy was almost a guaranteed by-product. The falsification of facts concerning new items from America was at a peak, and often it was not possible to distinguish between honest firsthand accounts and sensational ravings. Professional opinion of coca at this time wavered somewhere between Boerhaave's hint at the discovery of a genuine "elixir vitae" and the suspicion that another philosopher's stone hoax had emerged. People feared that the claims made on behalf of coca were simply untrue. Humboldt and Ruiz almost certainly discussed this topic and could not have helped but agree that it was the duty of science to lay this particular controversy to rest.

Humboldt knew before he left that the key to the solution of the coca question was to find its specific active ingredient. If he could positively identify the "bitter constituent" that Herman Boerhaave spoke of, he would help demonstrate that coca was possibly even beneficial and in turn advance the entire inquiry a giant step.

After all their affairs were settled in Spain, Humboldt and Bonpland booked passage on the sloop *Pizarro* and sailed out of Coruna on June 5, 1799. The trip across the ocean was practically uneventful; they stopped in the Canaries just long enough to avoid contact with a passing British squadron, then proceeded directly to Venezuela, landing there July 15 of the same year.

They immediately began to lay plans for what they called their Orinoco Mission. This mission involved a 700-league journey down the Rio Negro and into the Brazilian frontier, then a perilous trip to Cumana via Angostura. One evening, close to the departure date, Bonpland was attacked and almost killed by a huge *zambo* who became infuriated for some reason when he overheard Bonpland speaking French. Despite the attack, Humboldt and Bonpland began their journey on November 16, 1799, and returned as planned seven months later.

Encouraged by their early success, they now decided to go to Manila by way of Vera Cruz. Shortly after this trip began, they were overtaken and captured by pirates who were reluctant to allow the explorer to reach their planned destination. Fortunately, the British naval vessel *Hawk* in turn captured the pirate ship, took Humboldt and Bonpland aboard, and the two eventually ended up headed for Havana on an American warship.

While in Cuba, Humboldt learned that the French expedition he missed in the winter of 1799 with Baudin had left France and was expected to be in Peru within the year. The plans for traveling to Manila were canceled and Humboldt wrote to Baudin that he and Bonpland would depart Cuba immediately and would join the French party in Lima.

They left Havana on March 11 and did not stop until they reached the Peruvian province of Popayan, where they celebrated the Christmas of 1801. There Humboldt found the residents of the district using coca and after he learned that the natives always added a bit of lime in the process, it occurred to him that he may only have to look a little way beyond the leaf to learn the secret of its powers. He attempted to explain the coca phenomenon by speculating that it was the lime rather than the leaf that gave the user his exceptional powers.* This conveniently seemed

*Curiously, Humboldt made the same mistake regarding the betel nut.

to explain why many Europeans who tried coca without lime failed to obtain any positive results and were so disappointed with the drug's effect. His hypothesis may have been influenced by Dr. Hipolito Unanue, the first native Peruvian author to discuss coca. Unanue had written considerably about Indian history and customs and was well aware of coca and what it could do. He believed that the addition of lime to coca created a new, third property that was responsible for the plant's remarkable performance.

In spite of Humboldt's apparent naïveté, his confusion was understandable at this early date. In addition, his query regarding the source of potency in the coca leaf chewing process represents an important elaboration of Boerhaave's initial suggestion made a hundred years before. Even though Humboldt was wrong, his work brought the controversy closer to solution in this way; Humboldt was not merely ethnographic in his coca observations, he went beyond the descriptive level for the first time and focused attention on the following important question: What is the coca plant's source of potency; or simply, what makes it work? He assumed that the process of chewing coca with lime was a valid and effective method of combating fatigue, hunger, and pain and now he was implicity challenging nineteenth-century science to find the botanical base.

Many modern authors seize on Humboldt's observational error and discuss it as an unfortunate blunder made by an otherwise great man. Although his mistake is quite obvious to us now, it is important to remember that of all the Westerners who went to Peru and discussed coca, he was the first one to ask those important questions.

Humboldt and a few of his colleagues represented a league of distinguished European scientists who would no longer ignore native traditions concerning jungle plants. They were not conquistadors, priests, or pirates, but rather, the heroes of a new century of thought that had turned to science for solutions. They knew the materia medica of the Indians had genuine medicinal properties that would be of value to all men and at this point began to move in a direction that would lead to the isolation of the cocaine alkaloid eighty years later.

Baron von Humboldt and Bonpland left Peru in 1803 and returned to Europe via the United States. The details of their travels sparked considerable debate on many issues and prompted similar expeditions to return to the New World and pursue scientific explanations to the many problems generated by earlier explorations.

Not the least of these problems was the truth about the powers associated with coca-leaf chewing. Many people had been locked in ignorance by the Convention of Tordesillas, the document that gave Spain and Portugal exclusive travel rights in America. The reports of discovery and other New World developments made by Spanish priests and soldiers suffered the censorship of the Holy Office of the Inquisition, the Council of the Indies, and the Casa de Contratación.[12] Considering the enormous stumbling block represented by these three bureaucratic offices, it is surprising that so many fine reports and books were produced, but these texts were often suppressed and ended up collecting dust in basement archives in Seville and Madrid, where no one ever got to read them. Because of the Spanish restraint on information, word of coca barely leaked out and the subject remained shrouded in mystery. Consequently, many of the coca stories took on an excited tone, and exaggerated stances were assumed by those who favored the plant as well as those who did not. There was always some well-founded doubt that the rumors about the leaf were authentic, and it was suspected that coca might be an invention of the Spanish clergy.

Humboldt, Unanue, and Ulloa all offered enough information to suggest that there actually was something surrounding coca use that produced extraordinary results, but none of them was specific enough to resolve this controversial issue. Their reports did seem to end the mystery of coca's existence and also served as an inspiration for other men of science to travel to South America and take up the coca debate where they left off.

The men who followed Humboldt and others into the field had hardly begun their preparations when the coca controversy began to appear in local newspapers and popular magazines. By 1820 the coca question was being widely discussed in print and there was even a call for Professor Humphry Davey, the man who first discovered the analgesic properties of nitrous oxide, to rise to the occasion and proclaim that coca, too, was in the service of mankind.

An English subscriber to *Gentleman's Magazine and Historical Chronicle* wrote in 1817:

> While not yet fully acquainted with the secret with which the Incas sustain power, it is certain that they have that secret and put it in practice. They masticate coca and undergo the greatest fatigue without any injury to health or bodily vigor. They want neither butcher nor baker nor brewer, nor distiller, nor fuel, nor culinary utensils. Now, if Professor Davey will

apply his thoughts to the subject here given for his experiments, there are thousands even in this happy land who will pour their blessings upon him if he will be discover a temporary anti-famine, or substitute food free from all inconvenience of weight, bulk and expense, and by which any person might be enabled, like the Peruvian Indian to live and labor in health and good spirits for a month now and then without eating. It would be the greatest achievement—whatever a London alderman might think—ever attained by human wisdom. . . . Englishmen . . . must dwell with rapture upon the thought of the multitude of animals that would be spared from slaughter to supply the bloody habits of twelve million people were this Peruvian requirement adopted, only on alternate days, throughout the year.[13]

In the nineteenth century, there was ample opportunity for the scientist to reach South America and work with coca. Thanks to the political posture of La Condamine and Humboldt, the Spanish government came to regard the work of naturalists as harmless, and after the Napoleonic wars, South America entertained many shiploads of them from all over Europe and America.

The situation was particularly advantageous for the advancement of coca because it replaced the soldier with the plant scientist and shifted the emphasis from military or religious conquest to the identification and application of useful and previously unknown New World plants. Cinchona and coca were two new items that held outstanding promise in this category. They both attracted a tremendous amount of attention and became the subject of many investigations.

The Europeans gave up the struggle to resist cinchona in 1820 after two French chemists isolated the active ingredient, quinine; they followed the lead of Wilhelm Serturner, who separated morphine from crude opium sixteen years earlier. The coca question was still far from being resolved, however, even after these two important events took place.

One reason for this is that coca still was not fully accepted as an active plant. Next, for those who did believe in coca, there was a discussion surrounding the effects of the leaf on the human body. Some people believed that coca really did act to give strength and endurance and to end hunger, but only in the body of the Indian. Others saw the coca-chewing practice as ruinous, inventing fearsome addictions to go along with constant use and predicting death for those who dared to try.

It is understandable that the general public was slow in getting a realistic picture of this South American chewing custom. After Humboldt's return, there followed several decades of discussions and observations that were hallmarked by astonishing accounts of the plant, both pro and con. The works of von Tschudi and Poeppig serve as good contrasting examples of the type of confusing sensationalism brought to print in an attempt to clarify the situation.

Johan Jacob von Tschudi was a Swiss naturalist who visited Peru in 1838. After two years of traveling the countryside on horseback, making observations on a number of subjects, he descended into the *montaña* in the winter of 1840. In his accounts of the use of coca in this region, he tells of a sixty-two-year-old Indian by the name of Hatun Huamang who labored for him five days doing pick and shovel work, without food and with only two hours of sleep each night. After his fifth consecutive twenty-two-hour day of hard labor, von Tschudi reported that the man was still in condition to accompany him on an arduous twenty-five-mile journey. All along the way he reportedly jogged beside the mule that carried von Tschudi and did so relying wholly upon coca for his sustenance. Later, the Swiss naturalist spoke with another Indian, who claimed to be one hundred and forty-two years old; the man said that for the last ninety years he had drunk nothing but *chicha* and had chewed an ounce of coca three times a day since he was eleven. He added that he ate only modest amounts of corn, *quinoa,* and barley and had meat only once a week, on Sundays.

In light of this report, it is not surprising to learn that Johan von Tschudi was inspired to use the leaf himself, chewing it to sustain his respiration when traveling in the high country. In his writings he concluded that coca was in fact highly nutritious. In *Travels in Peru During the Years 1838–42,* he stated emphatically, "Setting aside all extravagent and visionary notions on the subject, I am clearly of the opinion that moderate use of coca is not merely innocuous, but that it may even be very conducive to health."[14]

On the other end of the spectrum was Edward Poeppig, a German naturalist who traveled extensively in Peru and Chile between 1827 and 1832. Poeppig was noted for his thoroughness, and made careful notes on the natural history of the region as well as on the customs and habits of the natives he encountered along the way. After he left Chile, Poeppig went to Lima, Cerro de Pasco, Huanuco, and finally into the *montaña,* where he was able to add the coca plant to his collection. It is clear from his accounts that he was not favorably inclined toward American aborigines or their ways. After his return to Europe, Poeppig's observations regarding the leaf and its effects on the Indians who used it were often quoted in widely read periodicals and even worked their way into standard reference works whose information was generally accepted as fact. Poeppig is quoted in the *American Cyclopedia* under the definition of *E. Coca* as saying, "The practice of chewing the leaf is attendant with the most pernicious consequences, producing an intoxication like that of opium. As indulgence is repeated the appetite for it increases and the power of resistance diminishes until at last death relieves the miserable victim."

The negative comments about coca infuriated von Tschudi and inspired him to search for the active agent in the leaves and send it back to Europe. Like Boerhaave and Humboldt, he realized that the discovery of the alkaloid would be the key to acceptance.

He made his intentions known to the director of the Laboratorio Botica y Droguería Boliviana in La Paz, a man by the name of Señor Pizzi. Pizzi received notice of von Tschudi's desires, along with a supply of leaves that were to be used in the separation experiments. After a while, von Tschudi received word from Pizzi that his experiments were successful and that he had managed to isolate the active ingredient. Delighted, von Tschudi took the substance back to Germany and personally placed it in the hands of one of the most celebrated pharmacologists of the day, Dr. Friedrich Wöhler of Göttingen, the man who synthesized urea. However, much to von Tschudi's disappointment, coca's battle for acceptance was thrown back an additional fifty years when Wohler announced that Señor Pizzi's preparation was nothing more than ordinary plaster of Paris. This report greatly added to the confusion, as it seemed to suggest to the general public that both von Tschudi and Poeppig were passing along tall tales. It made their comments appear ridiculous because it stirred up that old suspicion that coca was just an ordinary plant and that all its sensational qualities existed only in the imaginations of those strange little Indians or confused travelers.

Nevertheless, the coca controversy refused to die. Representatives from Bavaria, France, England, Russia, and even the United States continued to travel and study in South America, and many of them made it a point to pass comment on the much-debated Inca leaf. The misgivings surrounding the credibility of Indian herbals that resurfaced after von Tschudi's embarrassing experience with Wöhler would be forever laid to rest by the work of an English scientist who arrived in South America in 1849.

Richard Spruce is remembered as one of the greatest botanists of all time, and can be considered as the champion of New World drug plants. Born in Yorkshire in 1817, he made a pilgrimage to South America in his thirty-second year.

By 1851, Spruce found himself over a thousand miles deep in the dense jungles of the *Alto Orinoco.* He spent long days collecting plants under the worst possible conditions and managed to stay alive in spite of malaria, a diet of alligator meat, and at least one attempt on his life. In November of the following year, Spruce was gathering specimens along the Rio Uaupes unaware that later that evening he would experience perhaps the most unusual event in his already colorful life.

After a day of specimen collecting in the jungle, Richard Spruce came upon the central village of the area around sunset. The settlement, which consisted of mud and thatch huts, appeared unusually still and Spruce moved apprehensively towards the *moloka,* or community house. When he was just in front of the main hut, the prevailing silence was shattered by the sound of loud trumpets from the jungle. This was the signal that began the festival of the gods—a reckless twelve-hour adventure that featured all manner of intoxication and great circular dances involving hundreds of people. The village was suddenly filled with hordes of small brown natives rushing in every direction, dancing, drinking, and crashing into each other and Spruce with great abandon. As was his usual custom, Spruce took off his shoes and joined freely in the ceremony.

Before long, the English botanist was caught up in the crush of several dozen painted men wearing giant headdresses and strings of red seeds. They all danced over to a reserved area where there were gourds filled with a strange-looking liquid the men called *cappi.** A dancer picked up one of the gourds and proceeded to drain off all the brownish fluid in two great gulps. The hallucinogenic affects of the drink took effect after a few moments and sent the dancer into a wild delirium that lasted ten minutes, followed by a trancelike state that lasted several hours.

*Later identified (by Spruce himself) as *Banisteriopsis caapi.*

Spruce, who at this point was bound by protocol, was offered the drink next; he cautiously accepted and emptied half a gourd. During those first anxious moments when the drug started taking effect,* the eager Indians plied him with additional drafts of manioc beer, palm wine, and even encouraged him to puff away on a huge two-foot cigar brought for the occasion.

Spruce woke up the next day astonished in the aftermath of his first hallucinogenic experience. He was both impressed and surprised that there were indeed Indian plants that had dramatic, unmistakable effects on the mind and body. On the basis of this, Spruce decided to redirect the emphasis of his entire research and began to focus his energies on the world of narcotic plants and Indian materia medica.

In his subsequent investigations, Spruce discovered that coca, the energizer of the Peruvian *montaña,* had found its way down into the Amazon basin and was cultivated by Brazilian natives under the name *ipadu.* He noted that they prized the shrub for its medicinal effects and grew it in gardens around their homes.

Richard Spruce did for Indian narcotics what his friend Charles Darwin did for zoology. His work revolutionized the current thinking on an old topic and cast the problem in a new light. He made Europeans see Indian herbals not only as substances that actually worked but also as substances that could work for them as well. With this end in mind, he sent shipments of coca and other plants to Europe with complete references to their native names and their suggested application.

In spite of all his good work, Spruce died a pauper in a one-room cottage in England. It is gratifying, however, to learn that his new attitude toward Indian cures, together with some of his coca plants, reached Berlin and caught the interst of a young chemist named Albert Niemann, the man who would finally unlock the secret of coca by separating the alkaloid cocaine.

The effects of Spruce's work began to take hold immediately and influenced many medical men to initiate experiments with coca. Naturally, a great many of these investigations were aimed at the discovery of the active agent. In 1853, Dr. Weddell personally experimented with the leaves and recognized that coca was everything Spruce and the others said it was. He went on to guess that its sustaining effects were the result of

*It is debatable if Spruce actually consumed enough of the drink to produce the hallucionogenic effect.

**Coca leaves packaged in the traditional way.** ***(Courtesy of the Fitz Hugh Ludlow Memorial Library)***

the presence of theine, or caffeine, the recently discovered alkaloid in tea and coffee. One of Weddell's colleagues in Paris, Dr. Wackenroder, examined the leaves but found no trace of theine; instead, he obtained a peculiar tannin which he incorrectly declared to be the active agent.

Two years later, a chemist named Gaedcke published a coca article in *Archives de Pharmacie* in which he described a process that would yield the essence of coca. Gaedcke discovered that the distillation of a dry remnant of dilute coca leaves produced a strange-smelling oily liquor now known to us as hygrine. His refinement of this fluid gave him some small needle-like crystals that he called Erythroxyline, and we now call cocaine.

In spite of the fact that Gaedcke's researches demonstrated a truly active agent in coca for the first time, he is not remembered as the man who discovered cocaine or the one who put an end to the coca controversy. Gaedcke's work was not actually forgotten, but it became the object of professional skepticism when other chemists were unable to produce the same results from his formula. The inability to duplicate Gaedcke's experiments was not the result of any laboratory error on his part. The problem was that very few coca leaves were available to European chemists at the time, and even fewer fresh ones. The leaves they did manage to acquire were often brought back as botanical voucher specimens, and as such were unfit for chemical analysis. The preservation that takes place when a plant is pressed between the pages of a book is adequate if the goal is description, but to investigate the alkaloids properly, especially in a delicate plant such as coca, the specimens must be dried immediately and correctly packed before shipment. The handful of leaves that trickled into the chemistry labs of Europe in 1855 were poorly preserved; this fact helps explain why so many chemists found Gaedcke's experiments worthless and then went on to assume that the coca leaf was inert.

The necessity of using fresh leaves for experimentation was vividly demonstrated by Dr. Paolo Mantegazza, an eminent Italian neurologist who conducted experiements on himself while a resident in one of Peru's coca-growing regions. Given his location, it is reasonable to assume that Mantegazza used only fresh leaves for his experiments, but at the same time it is a bit unreasonable to believe that less than an ounce of those fresh leaves could produce the effects that Mantegazza describes. He claimed that a few seconds after chewing the leaves, "the phantasmagoric images came with such rapidity and the intoxification was so intense that I tried to describe the fullness of the happiness washing over me to a friend and colleague close by, but I spoke with such vehemence that he was not able to take down more than a few of the thousands of words with which I was deafening him." A few hours later, he said," I was sufficiently calm to write these words in a steady hand: 'I sneer at poor mortals condemned to live in this valley of tears while I, carried on the wings of two coca leaves went flying through the spaces of 77,438 worlds, each more splendid than the one before.' "[15]

When he returned to Italy, Mantegazza took a job in the Anthropology Department of the University of Florence and from that position continued with his coca publications. Despite or possibly because of his exaggerated reaction to the leaf and his extravagant prose, Mantegazza's first essay was very favorably received and is said to have inspired an era of scientific investigation into the nature of the coca shrubs' properties.[16]

Dr. Friedrich Wöhler, the pharmacist from Göttingen, who had previously worked with von Tschudi was one of the men who began to take a real interest in coca after the publication of Mantegazza's report. Wohler never forgot von Tschudi's enthusiasm when he had come to him a few years before, and probably remembered feeling a bit embarrassed when he had to inform the Swiss naturalist that he had been the victim of either an unfortunate mix-up or a cruel joke. He, of course, dismissed the possibility that plaster of Paris had anything to do with this New World plant, and could never get over the excitement and support given the shrub by so many eminent men such as von Tschudi.

After being alerted to the possibility that specially preserved or fresh leaves might yield the essence of coca's potency Wöhler personally contacted Dr. Karl Scherzer, a trade expert, and made arrangements with him to procure a considerale amount of the freshest Bolivian specimens. Dr. Scherzer was about to leave for a trip around the world aboard the Austrian frigate *Novara;* before he left, he made reference in his notes to Wohler's desires, quoting the pharmacologist as saying he wanted the leaves "to enable him to analyze more completely than had yet been done, the chemical constituents of this remarkable plant."[17] He also added that coca had "hitherto only reached Europe in small quantities, having in fact been carried home simply as curiosities."[18]

When the *Novara* crew reached Bolivia in 1859, Dr. Scherzer noted that the country's coca crop was just over 78,000 tons; he made it a point to set aside a little over a *cesto,* or about thirty pounds from this enormous amount, then carefully dried and packed the leaves and marked them for delivery to his friend Wohler as per their earlier discussions. Scherzer's *cesto* was probably the largest amount of coca ever to arrive in a European laboratory at any one time, and was certainly the largest amount of properly prepared leaves on the Continent.

Friedrich Wöhler was one of the most distinguished research pharmacologists of his day and simultaneously held a full-time teaching position at the local university. As a result of the latter, a great deal of his time was taken up with yeoman work for his department and many new projects had to be assumed by laboratory assistants.

When the carefully packed containers of coca leaves arrived from the *Novara,* Wöhler began the search for a graduate student to do the analysis. Fortunately for both of them, Herr Albert Niemann, the young student who had been so interested in the Richard Spruce coca specimens in Berlin, was perfectly predisposed for the job. He was a graduate student at the time, studying under Wöhler in his lab, interested in coca and also in search of a doctoral dissertation topic.

It was an ideal arrangement, and Niemann accepted the task without hesitation. He began to review the works of Wackenroder, Gaedcke, Weddell, and Unanue, everyone who helped to accumulate alkaloidal information on coca, and was undoubtedly already aware of the different travelers' reports and subsequent controversy. Working with properly cured, high-yield Bolivian specimens, he began his experiments by saturating some leaves in an eighty-five-percent alcohol wash containing a trace of sulfuric acid.

After he distilled off the alcohol, Niemann found that a syrupy mass was left, and from this he separated a resin. He then treated the resin with carbonate of soda and got an alkaloidal substance by repeatedly shaking it with ether. When he recovered the ether by distillation, he found an alkaloid present in proportion of approximately one-quarter of one percent, and he named this alkaloid "cocaine."

A few months afrer the *Novara* leaves appeared in Wöhler's lab, Niemann published *On a New Organic Base in the Coca Leaves,* his inaugural dissertation on attaining the degree of Doctor of Philosophy at Göttingen in March 1860. To obtain his Ph.D., Niemann refined the process of obtaining cocaine through a series of baths and distillations. Essentially, he rediscovered Gaedcke's "Erythroxyline" by means of an improved technique. Niemann's method was reliable and the cocaine he produced could be identified as being "in colorless transparent prisms, inodorous, soluble in seven hundred and four parts of water at 12°C, more readily soluble in alcohol, and freely so in ether. Its solutions have an alkaline reaction, a bitter taste, promote the flow of saliva and leave a peculiar numbness, followed by a sense of cold when applied to the tongue."[19] Niemann died a young man at twenty-six, only one year after the publication of his dissertation. He did not live long enough to realize that he had not only discovered the fugitive active alkaloid in the Inca leaf, but at the same time had narrowly missed the discovery of cocaine's greatest application. In his casual reference to cocaine's ability to produce a "peculiar numbness," Niemann overlooked its value as an anesthetic.

**Albert Niemann.** ***(Courtesy of the Fitz Hugh Ludlow Memorial Library)***

Dr. Niemann cannot be held responsible for missing the obvious about cocaine, for over the last 300 years only a scattered few had anything to say about the leaf's anesthetic effects. Even after Niemann first pinned down the major active agent and mentioned its numbing effect, the idea of applying cocaine as a local anesthetic would escape the combined forces of world research for another twenty-four years.

Niemann's cocaine lent scientific weight to the reports that the much-debated leaf was truly possessed of some remarkable properties and suggested that the stories the Incas told were true; there seemed to be an "elixir vitae" contained within the leaves, and it even arrived complete with the bitter taste that Herman Boerhaave predicted it would more than 150 years earlier.

The medical world had been stunned once again by an exotic item from the jungles of the New World. All at once several of these plants arrived on the scene to eradicate such ancient enemies as heart disease, malaria, and, of course, pain. Now Europeans had to prepare for the shock of realizing that still another plant had been "discovered" that would not only temporarily cure hunger, pain, and fatigue, but also actually seemed to give life.

Niemann's discovery brought on an era of physiological experimentation centered around both the coca leaf and its newly discovered alkaloid. It was at this point that the coca controversy shifted into an even more confusing state than before and one from which it has never fully recovered.

The confusion was generated by scientific experiments whose design was to clarify the subject but whose effects only confounded it. After learning of Niemann's discovery, a number of researchers made arrangements to acquire shipments of coca leaves from South America. Some investigators conducted their studies by having subjects chew the whole leaf or by having them drink mild infusions of coca that were also prepared from the whole leaf. On the other hand, some researchers naturally considered the new alkaloid the very essence of coca and elected to confine their investigation only to this most active ingredient and only after it had been isolated from the other parts of the whole leaf.

In the shadow of Niemann's great discovery, coca and cocaine became imperceptibly intertwined as items of contemporary interest. As a result, the difference between coca and cocaine in the minds of the general public was passed down as being almost nonexistent. The terms themselves became hopelessly mixed and, for some, almost synonymous. Niemann's recent revelation had suggested that cocaine was the sum of coca and *the* substance that was implied when speaking of the leaf's unique action. Because it was now certain that there was cocaine in every coca leaf, it was an easy thing to consider the two as one. When the test results of coca the leaf or cocaine the concentrate were added to this situation, all too often they were lumped together in a single category, with both layman and scientist neglecting to make the meaningful mental separation.

One of the earliest reports following Niemann's discovery came from an Austrian named Schroff. It was his intention to experiment with cocaine in hopes of coming up with a practical application for the new alkaloid. In one of his tests, he gave an oral dose of 0.05 grams to a puppy and noted some transitory mydriasis (excessive dilation of the pupil) and difficulty in breathing. When he applied the same amount subcutaneously, the puppy evidenced decided mydriasis, convulsions, and death. Later he injected the new alkaloid into the body of a healthy frog and discovered it had a severe paralytic effect on the animal's nervous system. In rabbits injected with cocaine, Schoff reported multiple spasms and convulsive death.

After people accepted cocaine's role in the reported action of coca, it was not difficult for them to interpret Schroff's experiments as supportive evidence for travelers such as Poeppig, who claimed that coca could kill you. Thus, the first experiment ever performed with cocaine only helped to denigrate the plant that produced it. Everyone seemed to ignore the fact that centuries of chewing whole coca had failed to produce even a single fatality. One year after Schroff made his work public, a German by the name of Dr. Fronmuller began to administer doses of cocaine to volunteers who were warned well in advance that the substance was probably toxic in some unknown amount. One of the more desperate volunteers seized on the warning and attempted to commit suicide by ingesting 1.5 grams of Fronmuller's cocaine. The German scientist reported that the man was foiled in his attempt because that amount of cocaine apparently had no effect on him. It should be noted that the amount of cocaine the volunteer allegedly ingested is more than the fatal dose listed in Goodman's and Gillman's *Pharmacological Basis of Therapeutics* and thirty times more than the amount that Schroff fed to the puppy.

In another strange case, Fronmuller gave an oral dose of 0.33 grams of cocaine to a man who had no preconceived notions one way or the other about coca or cocaine. Instead of acting like the "elixir vitae" of the supposed life-giving coca plant, Fronmuller's cocaine put the patient to sleep.

If the experiments carried out by Schroff and Fronmuller were not enough to confuse the coca issue thoroughly, then the more famous experiments of the next few years were certain to do the job.

Sir Robert Christison of Edinburgh was one of those who chose to conduct his experiments using whole coca leaves as the Andean Indians had done for centuries. In 1870, he carefully selected two students who guaranteed him that they had done absolutely no meaningful exercise in the past five months and then persuaded them to take a sixteen-mile hike after breakfast the next day. The students agreed and left early the following morning. Later that evening, they reported back to Christison completely fatigued and a bit out of sorts as a result of their professor's unusual request. Christison graciously received them and invited them into his home for refreshments. They were each given drinks containing two drachms of infusion of coca with five grains of carbonate of soda added to take the place of the lime used by the native chewers. After one of the more disgruntled students drank Christison's preparation, he unexpectedly burst out laughing. While the other was not so dramatically affected, they were both completely refreshed and were even moved to enjoy the most unlikely of all pastimes under the circumstances, an additional hour's walk. After the short hike, they settled down to a full meal and a sound, refreshing night's sleep.

The results of this experiment excited Christison and inspired him to continue with his investigations of the Peruvian leaf, this time using himself as the guinea pig. The next day, he chewed some leaves and hiked fifteen miles without taking food or water. Although not accustomed to any exercise, he claimed that he could not only proceed with ease but even "with elasticity." The test was repeated three days later with the same results.

The climax of Christison's experiments came six years later and involved a mountain-climbing adventure to the summit of Ben Vorlich, 3,200 feet above Loch Earn. After a rugged half-day of climbing, Christison's party stopped for lunch. While the other hikers eagerly filled their hungry stomachs, Sir Robert sat apart from the others and quietly chewed two-thirds of a drachm of coca leaves, his only food of the day. When lunch was over, Christison took the lead to the top and was clearly the most animated and least fatigued member of the group. Eight days later, when the other climbers were just beginning to recover from their aches and pains, Robert Christison announced that he would repeat the journey to the top of Ben Vorlich and do so using only thirty more grains of coca than he had used before and, of course, nothing else. Despite the fact that the weather had taken a nasty change and that it was a chilly forty degrees at the summit, Christison duplicated his feat with ease.

Christison's dual assault of Ben Vorlich while using coca was an impressive physiological statement and speaks for itself. His experiments were of significant value in terms of publicity because he was, at the time, the president of the British Medical Association and a well-known professional figure. Probably the single most dramatic item associated with the entire episode, however, was the fact that Sir Robert performed these physical gymnastics when he was seventy-eight years old!

Professor Christison summed up his experiments by saying, "I feel that without details the general results which may now be summarized would scarcely carry conviction with them. They are the following: The chewing of coca not only removes extreme fatigue, but prevents it. Hunger and thirst are suspended but eventually appetite and digestion are unaffected. No injury whatever is sustained at the time or subsequently in occasional trials."[20]

Only a few years before Christison's experiments were printed in the *British Medical Journal,* the former chief surgeon of the Peruvian army, Thomas Moreno y Maiz, conducted some experiments of his own. Using cocaine rather than the whole leaf, he tried to determine if coca could remove hunger and thirst and serve man as a temporary food replacement in emergencies. For his experiments he chose a dozen healthy rats and divided them equally in two separate cages; one group was not given any food or water and the other group was only given the alkaloid cocaine. If what the Peruvian Indians and others claimed was true, the first group would be expected to die of starvation while those given only cocaine would, for a time, be spared. After a few weeks the experiment ended and the results were unmistakable: the rats given cocaine died even sooner than those given nothing!

What Moreno y Maiz succeeded in demonstrating was not that coca and cocaine are ineffective food substitutes or fatigue-relieving agents, as his experiments suggested, but rather that a steady diet of undiluted cocaine alkaloid and nothing else might kill you faster than starvation. Had he given the second group infusions of the whole leaf instead of injections of only the concentrate, the rats in the second cage may have indeed survived longer, for the amount of thiamine, riboflavin, and vitamin C in whole coca is almost enough to satisfy the minimum daily vitamin requirements of an adult Peruvian Indian.[21]

Unfortunately, the conclusions drawn from the Moreno y Maiz cocaine experiments were taken as additional proof that the claims made about coca's sustaining powers were nonsense and, if anything, only complicated an already confused situation. To make matters worse, an English doctor by the name of G. F. Dowdeswell acquired some poorly preserved, inert leaves in the same year as Christison and said, "[I used coca] . . . in all forms, solid, liquid, hot and cold, at all hours, from seven in the morning until one or two at night, fasting and after eating, in the course of a month probably consuming a pound of leaves without producing any decided effect. . . . It occasioned not the slightest excitement, nor even the feeling of buoyance and exhilaration which is experienced from mountain air or a draught of spring water."[22] Possibly referring to the septuagenarian Christison, he added, "The subjective effects asserted may be curious nervous idiosyncrasies."[23]

In an attempt to refute Dowdeswell on the basis of relative freshness and potency of the leaves used in the two experiments, Christison was forced to admit that the ones he used were "at least seven years old," and probably older than Dowdeswell's. On the other hand, he correctly pointed out that, although they were old, his leaves were still green, flat, unbroken, bitter to the taste, and full of aroma, restating the probability that carefully dried and packed leaves could keep their potency for years.

This was apparently a little too much for the English people to swallow. Many lined up behind Dr. Dowdeswell, who claimed that coca was inert and produced no results; others forgot about coca altogether and wondered if cocaine had any reliable action. A third group feared if either substance had a meaningful effect on the human body, that effect would be to kill it!

Sir Robert Christison's efforts on behalf of coca were dismissed and England displayed a certain naïveté in this matter by assuming a strong prejudice against the leaf. The situation was not likely to change in England either, for as a result of bad publicity, it became even more difficult to obtain any leaves for additional experiments. In 1870, British physician John Bennett found that it was next to impossible to obtain two pounds of coca.

Although most of England came to believe that coca was either worthless or fatal and made no efforts to deal with it at home, it is clear that some enterprising concerns in the United Kingdom decided not to take any chances and did some agricultural experiments in the Commonwealth nations. Anticipating that coca could possibly have some marketable value, and quite oblivious to their own suspicions that it had great potential danger, the Royal Botanical Gardens at Kew distributed coca to every crown colony that could support its growth.

English ships carried the leaf to Ceylon in 1870 and later cultivated it on the Indian subcontinent in the district of Nilgiris. Other plants were shipped to English possessions in Africa and on the Malay peninsula.*

Taking its cue from British merchants, the enterprising Dutch East India Company introduced the leaf to Java in 1870, where it later achieved great success. A communication made by Dr. Betances to the Société d'Acclimatation de France suggested that even Spain and representatives of the Catholic Church were now attempting to distribute the plant they had formerly tried so hard to suppress. In his report, Betances stated that at considerable expense and after numerous shipments of seeds and the transportation of seedlings to Puerto Rico and the Dominican Republic, he had the pleasure of receiving a fine branch of coca in full bloom. He added that the specimen was sent by Monsignor Mereno, the archbishop of Santo Domingo.[24]

The efforts to distribute coca around the world were motivated only by a suspicion that it might have some practical value. The first European experiments with both coca and cocaine led many people to entertain serious misgivings about the plant's effectiveness either as a stimulant or as an anti-famine device, and with these two avenues of application in doubt, the future of either substance as an item of practical value seemed dim. In most places, even ardent supporters of the leaf were hard-pressed to come up with any alternate uses for it.

The debate to establish cocaine's effectiveness as a bracer or food substitute not only kept interest sidetracked following the results of confusing experiments, but it also obscured some early statements that unknowingly but accurately predicted cocaine's most important application. It is interesting to learn that a number of these early predictions were contained in some of the very scientific experiments that helped generate the debate in the first place.

Professor Schroff's cocaine poisoning of the puppies completely overshadowed his passing reference to the fact that the alkaloid had the ability to numb the tongue. Moreno y Maiz's comments were even more explicit, but equally unheralded. At the end of his much-discussed paper on cocaine, rats, and starvation, he put some cocaine into his mouth and noticed, like Schroff, that his tongue had become anesthetized. Afterwards, he asked, "Could one use cocaine as a local anesthetic?" Unfortunately he immediately backed off and answered himself by saying, "It must be decided by the future." Another forgotten aspect of Moreno y Maiz's paper is his early recognition of cocaine's famous euphoric effect. When he accidentally anesthetized his tongue, Moreno y Maiz commented on his disposition shortly after that event took place, saying, "These were the most blessed moments of my life."

The names of Schroff and Moreno y Maiz could be added to a long list of other scientists who made incidental discoveries about cocaine: Wöhler; Niemann; Demarle; the Russians Nikolsky, Tarkhanov, and Danni; von Anrep and Sigmund Freud were some of the most famous. They all wondered out loud about cocaine and local anesthesia, but none of them was able to make the two stick together, somehow failing to recognize the significance of their own observations.

*The descendants of these original plants are still grown as ornamentals in conservatories in North America and Western Europe as well as in India, Sri Lanka, Indonesia, Hong Kong, Jamaica, and Zaire.

# 8

# Freud, Koller, and the Discovery of Local Anesthesia

While most of Europe was occupied by the debate over the effectiveness of cocaine, a Corsican chemist by the name of Angelo Mariani was putting together the first modern industry based on coca.

Mariani was born in Batia, the largest city in Corsica, and was raised in a family of physicians and chemists. Around the time of Niemann's discovery, he moved to Paris and became interested in securing some coca and experimenting with it. Sometime after Mariani first used the leaf, he realized that there was no mistaking its unique action; as far as he was concerned, the great coca debate had come to an end. Mariani was certainly not oblivious to the heated controversy that continued to surround coca and cocaine, but he did not allow that controversy to keep him from his convictions or ambitions. Angelo Mariani decided that he would capitalize on his belief in the plant's abilities and make this substance available to anyone who had five francs in his pocket to spend on it.

Mariani built his empire on a solid foundation. Before he decided to make coca available to the public, he gathered travelers' reports from the New World and studied them carefully, paying close attention to the ones that described the selection process that takes place when a highland Indian buys his coca. He was especially interested in the fact that the more bitter leaves were rejected by many Indians even though it was recognized that they were the ones that contained the highest percentage of cocaine. He also became familiar with the many different varieties of coca and the particular aromatic qualities for which each was supposed to be famous.

In addition to being a chemist and a very perceptive coca scholar, Angelo Mariani was also a shrewd businessman. He was aware that coca would have to be marketed in a very positive way and that it would have to be made available in some new form that would be both familiar and acceptable to nineteenth-century Europeans. By this time, it was obvious that Westerners were not inclined to chew it.

To achieve this, Mariani ordered several different varieties of leaf and began blending them together in a unique way, taking great care to set aside the bitter ones just as the Indians did. He found that by steeping this blend in fine Bordeaux wine he could preserve both the flavor and action of whole coca as well as present the product in a form that most people felt very comfortable with. He called his product Vin Mariani and advertised it as an unequaled tonic-stimulant for the fatigued or overworked body and brain.

Mariani became the world's largest importer of coca leaves almost as soon as he opened his doors in 1863. Encouraged by his success with the wine, he soon expanded his operation to include lozenges, elixirs, tea, and pâté, all made by

**Angelo Mariani.** ***(Courtesy of the Fitz Hugh Ludlow Memorial Library)***

him from the whole leaf and never produced by the simple addition of cocaine.

Two years after Mariani produced his first bottle of coca wine, he was approached by Dr. Charles Fauvell, a Paris laryngolist, and asked to make a special coca preparation for use in the doctor's clinic. Fauvell had an idea that coca may affect the mucous membrane of the larynx in such a way as to relieve the throat pain often experienced by professional singers and public speakers. He also suspected its importance as a tensor of the vocal chords.

Mariani was more than eager to lend a hand in support of the therapeutic use of coca and he quickly filled Fauvell's prescription. Armed with the Mariani preparation and a good working hypothesis, Fauvell began making local applications of coca for the purpose of anesthesia. His success in treating the singers of Paris became well known in the local music profession and because of his work, French vocal instructors of the day had their pupils sip Mariani's wine during rehearsals as well as recitals.

While Fauvell's idea found favor with hundreds of opera stars and stage sopranos, his medical support for the first modern application of local anesthesia numbered only three. Dr. Morell Mackenzie and Dr. Lennox Browne of England, as well as Dr. Louis Elsberg of America, traveled to Paris to observe Fauvell's clinical successes firsthand. They were so impressed that they began to use coca in their own practices, and each became a supporter as well as imitator of Charles Fauvell.

Fauvell's therapeutic application of Mariani's wine as a local anesthetic should have rocked the medical world and marked a turning point in coca's battle for acceptance. His experiments were premeditated and successful and there was no question about his appreciation of the significance of his work or the fact that he put coca and local anesthesia together and made

**Advertisement for Vin Mariani.** ***(Courtesy of the Fitz Hugh Ludlow Memorial Library)***

them stick. If it were not for the blinding controversy that surrounded the plant at that time, Fauvell would probably be remembered as a famous man and would have made all subsequent references to coca, cocaine, and local anesthesia rhetorical.

Looking back on the episode twenty-five years later, Angelo Mariani wrote: "Charles Fauvell was the first to make use of it [coca] as a general tonic, having a special action on the larynx; and to make known its anesthetic and analgesic qualities. . . . Although striking effects were obtained from this valuable medicine, its full worth was as yet unknown and there was diversity of opinion as to its mode of action."[1]

While Fauvell never became famous for his efforts, his work did play an important role in the development of coca as a useful agent. His two students Morell Mackenzie and Lennox Browne returned to England, where they enthusiastically attempted to follow up on the Paris experiments. Unfortunately, their work was ignored in a part of Europe that had already decided against the leaf. The third protégé, however, Dr. Louis Elsberg, carried Fauvell's idea as well as Mariani's wine back to the United States, where they were both free to develop on their own merits and apart from the controversy that was stifling European thinking.

The idea of using coca as a therapeutic tool fell on fertile ground in the United States and it was there that the leaf took a giant step toward acceptance. The social climate and the state of medicine in America were the major factors in this process.

When Elsburg returned to the United States, he came back to a country where there had always been a shortage of doctors and a great reliance on home cures and medicines. Americans were thrown on their own resources then and the subject of how to get well was a wide-open question. Understandably, frontier Americans were heavily influenced by the medicine of the many aboriginal tribes they encountered wherever they went and, by all accounts, medicinal root and herb use by the Indians in the pre-Columbian North America was very extensive.[2] Because Americans were already using drug plants and claiming success with them, an atmosphere of acceptance for botanical remedies was already established. White settlers in America could see firsthand what their European cousins could only be told about. As a result of this, it was easier for them to accept new and different drug cures, especially ones from the plant world.

This effect began to manifest itself as early as the 1820s, when a self-taught physician from New Hampshire named Samuel Thomson embraced the Indians' love of plant cures and published a book called *New Guide to Health or the Botanic Family Physician.* Basically, Thomson believed there were only two ways to cure a sick patient; one was by a bizarre process he called "external steaming" and the other was by the therapeutic application of remedies from the plant world. In 1832, he established the Friendly Botanic Society and six years later the Botanico-Medical College of Ohio. Thomson commanded a large group of receptive Americans until splinter groups formed within his own discipline and sent the New England physician into obscurity.

**Charles Fauvell. *(Courtesy of the Fitz Hugh Ludlow Memorial Library)***

In the early 1860s, Thomsonian medicine in America was replaced by homeopathy and the followers of Herr Samuel Hahneman of Germany. One of Hahneman's basic propositions was his belief in a spiritual healing power that was locked up inside plants.

On the more conventional side of the plant-cure issue was the prominent American botanist C. S. Rafinesque, who prepared the text called *Medical Flora; or Manual of Medical Botany of the United States of North America.* Rafinesque urged Americans to research plant chemistry and "adopt in practice whatever is found beneficial."[3]

In light of this long-standing relationship be-

The Inca city of Machupicchu, formerly known as Vilcapampa, was a religious center and surely the locus of coca use and probably cultivation. Photo by Joseph Kennedy. *(Courtesy of John A. P. Kruse)*

An Indian *brujo* prepares a sacrificial bundle containing coca leaves. Photo by Loren McIntyre. *(Courtesy of Loren McIntyre)*

Golden Inca figurine with bulging cheek. Photo by Loren McIntyre. *(Courtesy of Loren McIntyre)*

**Coca, together with other plant goods from the Andes. Photo by James Powell.** ***(Courtesy of James Powell)***

**Peruvian Indian with a coca leaf applied as a poultice for a headache. Photo by Loren McIntyre.** ***(Courtesy of Loren McIntyre)***

**An American advertisement for Vin Mariani. Photo by Arturo Wesley.** ***(Courtesy of Union Centrale des Arts Décoratifs)***

**Fresh coca leaves in a Peruvian market. Photo by James Powell.**

**A label for a bottle of Inca Wine drawn by the artist Mucha. Photo by Arturo Wesley. *(Courtesy of Union Centrale des Arts Décoratifs)***

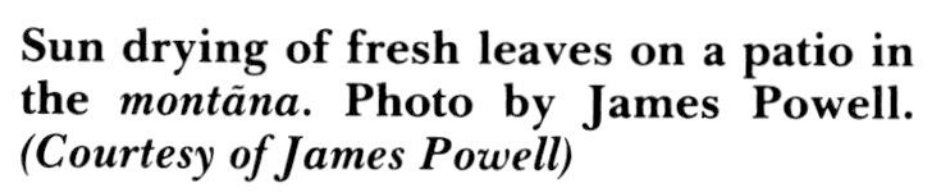

**Sun drying of fresh leaves on a patio in the *montãna*. Photo by James Powell. *(Courtesy of James Powell)***

**Sun drying of fresh leaves on a patio in the *montaña*. Photo by James Powell. *(Courtesy of James Powell)***

**"Garçon un Coca des Incas." Photo by Arturo Wesley.** ***(Courtesy of Union Centrale des Arts Décoratifs)***

**Man holding a bag of coca in doorway. Photo by James Powell.** ***(Courtesy of James Powell)***

**Indian family tends coca terraces in Bolivia. Photo by Loren McIntyre.** ***(Courtesy of Loren McIntyre)***

tween Americans and botanic-based cures, it is not surprising to learn that Mariani's wine realized its greatest popularity in the United States.[4] By 1880, bottles of Vin Mariani filled the shelves of American apothecaries and grocery stores and in the same year, the plant that it was made from was officially recognized and listed in the *United States Pharmacopoeia.*

Although it enjoyed widespread professional and popular acceptance, the coca plant did not arrive in the United States entirely unassailed. Dr. G. A. Ward, an American physician who had lived in Peru, cautioned: "I have been conversant with coca for the past eight years and would say that the great encomiums lavished upon its use are all 'bosh.' Coca will relieve and in a measure prevent thirst but so will chewing a bullet. I do not think it will prevent hunger as well as tobacco. In the meantime I would only advise my professional brethren not to consider that the elixir of life has been found in the leaves of E. Coca."[5]

Criticisms of this kind never took root in the United States and the popularity of coca continued to grow. Despite Ward's admonition, it was his "professional brethren" who were especially pleased with the leaf's beneficial effects and were largely responsible for its popularity and distribution throughout the country.

American physicians were not slow in recommending coca in all forms to their patients and did not shrink from experimenting with it for a number of diseases. They prescribed coca for dyspepsia, depression, hypochondria, cardiac weakness and cardiac irregularities, hysteria, stomatatis, melancholia, asthma, hay fever, stupor, cachexia, and as an aid to rid the blood of uric acid; it was also used as an aphrodisiac and, of course, a mild tonic stimulant. The results of many of these experimental applications were published in the pages of the *Detroit Therapeutic Gazette,* a monthly journal that reflected coca's acceptance and popularity in the United States by running a total of sixteen articles on the plant and its possible applications.

One of the contributions to the September 15, 1880, issue of the *Therapeutic Gazette* created a considerable stir because it promised the first antidote for two of man's most severe afflictions. Dr. W. H. Bentley of Oak Valley, Kentucky, wrote an article entitled "Erythroxoyn Coca and the Opium and Alcohol Habits", wherein he lists the cases of a half dozen patients who had been relieved of their habit by drinking coca preparations.

In July of 1878, Bentley was called to a house somewhere in the hills of Kentucky; when he arrived there, he was astonished to find a tenth of an acre in opium poppies *(Papaver somniferium).* The lady of the house, a forty-year-old widow, confessed to the doctor that she had been an opium eater for quite some time, consuming about a half pound of the drug a year. Bentley says: "I persuaded her to give up the habit. She declared that she could not. She agreed, however, to try, so I sent her one pound of the fluid extract of coca to begin with. When used up, she sent for half the quantity, stating that she thought it would complete the cure. I sent her one half pound. She sent me her opium crop that winter, with a message that the medicine had cured her."[6]

The medicine Bentley prescribed for her was the fluid extract of coca manufactured by Parke Davis and Company in Detroit, Michigan. The preparation they offered was standardized in the *United States Pharmacopoeia,* 6th to 8th editions (1880–90), and contained 0.5 grams of coca alkaloids per one hundred cubic centimeters of solution.[7] Of course, cocaine was one of those alkaloids contained in the fluid extract, but according to Bentley's recommended dose (1 drachm of fluid extract when the desire for whiskey or opium is quite urgent), a patient would only consume .0185 grams of cocaine per dose, and even then it would automatically be taken only in conjunction with coca's natural complement of other alkaloids; because it was prescribed to be drunk, this solution would also necessarily have to pass through the liver and kidneys. In these two important respects the use of the fluid extract closely resembled the Indian method of taking the drug, a procedure proven harmless by centuries of use. It is also interesting to note that one would have to take 13.5 one-drachm doses of Parke Davis fluid extract just to get the amount of cocaine alkaloid consumed daily by an average Peruvian Indian. This is based on Hanna's study that estimates the average Indian digests approximately 0.25 grams of cocaine a day by normal coca use.[8]

The American people were anxious to entertain any serious talk of a cure for morphine addiction because of the great number of Civil War veterans in the country who had accidentally been made slaves to the habit. Untrained field medics and hospital workers freely dispensed morphine to wounded soldiers on both sides and, as a result, the United States had an unusually large number of postwar addicts. By 1870, morphine or opium addiction became known as "Army disease."

Bentley's idea of curing these addicts with coca caught on immediately and was put into practice throughout the country. Reports of its effectiveness began to appear as a regular feature in the *Therapeutic Gazette* as well as the *Philadelphia Medical Times, New Remedies, Medical and Surgical Reporter,* and *New York Medical Record.*

The Americans had popularized coca. An editorial in the *Louisville Medical News* seems to have captured the national attitude toward the plant when it said, "One feels like trying coca with or without the opium habit. A harmless remedy for the blues is imperial."[9] The willingness of American physicians to prescribe the drug against a variety of diseases had led them to a spectacular application that could not possibly be ignored. The subsequent applause for coca in the New World was so great that the Europeans were forced temporarily to interrupt their great coca debate and take notice. The result was that coca was reintroduced to the Continent in a most positive way and this, in turn, seemed to be enough to break the spell of confusion that had many people going around in circles.

Following the notoriety coca received in America, two interesting articles appeared in Europe that seemed to reflect a new attitude toward the plant. Professor von Anrep and Dr. Theodor Aschenbrandt, both of Würzburg, published important experimental studies that suggested the American application of coca might definitely have some sound medicinal value and be worthy of additional investigation. Their work served as a rallying point for coca supporters everywhere and reawakened interest in the plant among those Europeans who might have otherwise forgotten about it.

In 1880, Professor von Anrep initiated a study of the effects of cocaine on animals. At the time, he was an assistant to Professor Rossbach at the Pharmacological Institute of Würzburg and was anxious to rework some of Dr. Schroff's ideas in light of the recent American announcements. Von Anrep's work produced an instant point of contention to Schroff's study because he treated animals for thirty days with moderate doses of cocaine and detected no deleterious effects on their bodily functions.[10] Schroff, it will be remembered, reported that the drug caused a horrible convulsive death in puppies, frogs, and rabbits.

Von Anrep went on to announce that even after taking large doses, warm-blooded animals only registered a strong psychic agitation and an excitation of the brain centers that control voluntary movements. Like Schroff, von Anrep administered the drug to dogs in low doses, but unlike Schroff's, his results showed only that the experimental animals had an initial rush of happy excitement followed by a manic compulsion to move. These findings were considered most important because they were directly analogous to Dr. Mantegazza's report on the effect of coca in the human body. From these observations, von Anrep concluded that, first and foremost, cocaine was indeed a great stimulant to the central nervous system; he then mentioned other effects of the alkaloid, which included accelerated respiration, a great increase in blood pressure, dilation of the pupils, and a speeding up of intestinal movement. Even after he administered enough cocaine to paralyze the respiratory center of a dog and kill it, he noticed that the striated muscle substance of the animal remained intact.

Von Anrep concluded his report with two general statements about cocaine that amounted to the greatest single concentration of insightful information on the drug so far. First, he made it a point to mention that cocaine and coca were two different substances and each had its own action; next, he described the locally numbing effect of cocaine and reiterated his belief that this drug might someday become a matter of medical importance.

In 1883, another assistant at the Pharmacological Institute, Dr. Theodor Aschenbrandt, picked up von Anrep's publication and became very interested in the medical possibilities his colleague mentioned at the end of his paper. Aschenbrandt also noticed that von Anrep had intended to follow up his experiments using humans instead of animals but had never found the time.

Aschenbrandt apparently did have the time and decided "to take up and continue von Anrep's work . . . with human beings. My study is intended to demonstrate that the alkaloid of the coca leaf, cocaine, is the substance that possesses the 'miraculous' quality described by Mantegazza, Moreno y Maiz, Dr. Unanue, von Tschudi, etc."[11]

Dr. Aschenbrandt began searching for an appropriate situation to make his tests. He needed "a mass of healthy people, exertion of all kinds, dry, hot weather and above all, the possibility of administering cocaine without anyone being aware of the fact that he was being observed."[12] This opportunity presented itself unexpectedly when Dr. Aschenbrandt was called to serve his reserve training unit in the Bavarian army.

The 1883 fall maneuvers of the second artillery battalion gave Aschenbrandt the chance to administer cocaine to soldiers for a variety of complaints. An exhausted man who had collapsed on the second day of the march was given a tablespoon of water with twenty drops of cocaine in solution. After waiting about five minutes, the man stood up on his own accord, put his backpack on and "cheerfully" marched the rest of the way with the other men. In another case Aschenbrandt spiked his servant's coffee with 0.01 grams of cocaine after the man complained of being poisoned the night before. The servant later told Aschenbrandt that he woke up that morning feeling even sicker than the night he was poisoned, but after taking his morning coffee, the dizziness and nausea seemed to disappear and he was able to rejoin his comrades. This particular case seems to have influenced the doctor most directly. After a restless night's sleep, Aschenbrandt woke up feeling bad: "I was really freezing, my head was not together and the prospect of the coming maneuvers was not something to cheer me up."[13] Although he was not really sick in any way, Aschenbrandt elected to put some cocaine in his own coffee just to drive away his early-morning depression, and found that the drug worked equally well for him.

Aschenbrandt concluded his report by saying, "I hope that with this study, which certainly is not complete nor entirely exact as to dosage and which certainly does not claim to be final proof of the properties of cocaine, I have drawn the attention of the military and inspired them to further research. I believe I have given sufficient evidence of its eminent usefulness."

Aschenbrandt's paper enjoyed only a very limited circulation and did not seem to stimulate any further military research.[14] However, at least one copy of his paper did find its way to Vienna and into the hands of Dr. Sigmund Freud, who became very interested in pursuing the therapeutic possibilities of the coca plant and its derivatives.

In early 1884, Freud was a young, ambitious doctor working in the famous General Hospital in Vienna. From the recent publication of many of his personal letters, it is clear that Freud had a great desire to be successful and was always alert to new medical possibilities that might bring him acclaim and professional advancement. He felt if he could make some sort of scientific discovery, his position in the hospital would be secure, he would have enough money to get married, and his reputation as a man of science would be established beyond a shadow of a doubt. His enthusiasm along these lines had already inspired him to invent the gold-chloride method of staining nerve tissue, which had given him the confidence to continue his search for additional medical revelations.

On April 21, 1884, Freud wrote a letter to his fiancée, Martha Bernays, briefly oultining his latest idea:

> I am also toying with a project and a hope which I will tell you about; perhaps nothing will come of this either. It is a therapeutic experiment. I have been reading about cocaine the effective ingredient of coca leaves, which some Indian tribes chew in order to make themselves resistant to privation and fatigue. A German* has tested this stuff on soldiers and reported that it has really rendered them strong and capable of endurance. I have now ordered some of it and for obvious reasons am going to try it out on cases of heart disease then on nervous exhaustion, particularly in the awful condition following the withdrawal of morphine. There may already be any number of people experimenting on it . . . ; perhaps it will not work. But I am certainly going to try it and, as you know, if one tries something long enough and goes on wanting it, one day it may succeed. We need no more than one stroke of luck of this kind to consider setting up house."[15]

Freud used Aschenbrandt's source and ordered a gram of cocaine from Merck of Darmstadt. When the drug arrived, he was outraged to learn that the bill for this small amount came to three gulden, thirty-three kreuger, or about $1.27. He had expected to pay only 13¢. The high cost almost persuaded the unhappy Freud to send the product back to Darmstadt and cancel his research, but fortunately he changed his mind rationalizing that he would be able to pay Merck Chemical Company sometime in the future. He accepted the drug and immediately tried the effects of a twentieth of a gram. After a few moments Freud realized that his mood of outrage had taken a dramatic change for the better and he had the feeling "that there is nothing at all one need worry about."[16]

This effect of cocaine use came as an unexpected surprise to Freud. His excitement about this new aspect of the drug seemed justified for he realized that he had just self-administered a remedy that could have some positive effect against a number of neurotic problems that had

*Theodor Aschenbrandt.

been plaguing both him and some of his friends. One month after first using cocaine, Freud wrote to his fiancée, saying, "I take small amounts of it regularly against depression and against indigestion and with the most brilliant success."[17]

Freud's enthusiasm for cocaine continued to grow. He had successfully ended the suffering of a man who had gastric catarrh with cocaine, and he sent the drug to Martha Bernays to "make her strong and give her cheeks a rosy color." He began to refer to cocaine as a "wonder drug" and started sharing his purchase with his colleagues, prescribing it to his patients, and mailing it to many of his relatives. Of this period in Freud's life, Dr. Ernest Jones, Freud's personally selected biographer, has remarked: "From the vantage point of our present knowledge, he was rapidly becoming a public menace."[18]

By this time Freud had gotten hold of Bentley's article in the *Therapeutic Gazette* and had decided to use cocaine to cure his friend Dr. Ernst von Fleischl-Marxow of his morphine habit. Fleischl, possibly the most gifted of all the great medical scientists associated with the General Hospital at this time, was an intellectual model for his friend Sigmund Freud. He had become addicted to morphine while attempting to battle neuromata, a painful disease of the nervous system he had contracted while doing research in pathological anatomy. Freud shared his initial gram of cocaine with Fleischl and after awhile began injecting it subcutaneously to his friend against pain.[19] Fleischl immediately embraced the drug and began to use it almost continually. This first administration took place in May of 1884, and for a brief time thereafter the treatment seemed very successful. Fleischl had apparently become the first morphine addict in Europe to be cured by the use of cocaine.

**Sigmund Freud in the late 1880s.** ***(Courtesy of Mrs. Hortense Koller Becker)***

These early experiences led Freud to believe that he had at last found the answer to all his problems. Shortly after he had given the cocaine to Fleischl, he wrote another letter to Martha Bernays, saying, "If it goes well I will write an essay on it and I expect it will win its place in therapeutics by the side of morphine and superior to it. . . . I hope it will abolish the most intractable vomiting, even when this is due to severe pain; in short it is only now I feel I am a doctor, since I have helped one patient and hope to help more. If things go on in this way we need have no concern about being able to come together and stay in Vienna."[20] Eight days later, he wrote back, saying that he was busy collecting literature for a "song of praise for this magical substance."[21]

In the beginning of his research, Freud had a great deal of difficulty obtaining printed material on coca. When Fleischl learned of his friend's problem, he wrote him a letter of introduction to the Society of Physicians, and it was in the society's library that Freud found what he needed. *The Index-Catalogue of the Library of the Surgeon-General's Office of the United States Army* published a complete bibliography on the subject of *E. coca.* With this set of references in hand, Freud was prepared to assemble one of the best essays on the subject ever written. While in the process of preparing the manuscript, Freud was also given permission to use the personal library of Dr. August Vogl, a Viennese pharmacist who for years had been quietly using a cocaine tea he brewed himself.[22]

As Freud began pouring through the coca literature, he was confronted with the possibility that the controversy over the drug's action could have been a function of subjective reaction on the part of different experimenters. This might have explained why many researchers rejected the drug, thinking it did not work. Before Freud could continue with his essay on coca, he also wanted to be absolutely sure that the drug's famous powers could be objectively verified and that they were not just invented by individuals paying lip service to coca's reputation.

In order to satisfy his curiosity, Freud set up a series of tests with a dynamometer (a device used to investigate the motor power of certain muscle groups) and called on a friend of his, Carl Koller, to help with the experiments. Koller later recalled the tests in this way: "In the summer of 1885, Freud, who had been interested in the physiologic systemic effects of cocaine, asked me to undertake with him a series of experiments in that direction. So Freud and I used to take the alkaloid internally by mouth, and after the proper length of time for its getting into the circulation we would conduct experiments on our muscular strength, fatigue and the like."[23] Freud also called on Dr. Leopold Konigstein to experiment with coca and urged him to "test the pain-deadening and secretion-shrinking effects of cocaine on the diseased conditions of the eye."[24] The results of the dynamometer tests alone were enough to convince Freud to continue with his song of praise to coca. With what he considered experimental evidence in hand, Freud could now take a step back and see the phenomenon of the coca controversy as an example of neglect through slander and an opportunity for his own advancement almost too good to be true.

After the dynamometer experiments, Freud eagerly resumed work on his manuscript; his aim was to reintroduce coca and popularize its therapeutic use. To achieve this, he arranged all the known facts about the leaf into a package that began with a discussion of the plant itself and then followed the history of the drug through two continents and a thousand years of association with man.

Freud's style in this monograph was a departure from the usual medical-journal format. Throughout the text, he took an editorial approach towards the subject matter and selected words and phrases that gave the article a strong pro-coca flavor. He stated, for instance, that he was inclined to accept Mantegazza's allegations even though he had not had an opportunity to confirm them and referred to the earlier European experiments with coca as leading to "a great disillusionment."

Freud called his article *Über Coca* and divided it into six parts: "The Plant"; "The History and Use of Coca in Its Country of Origin"; "Coca Leaves in Europe—Cocaine"; "The Effect of Coca on Animals"; "The Effect of Coca on the Healthy Human Body"; and "The Therapeutic Uses of Coca." The last section contained recommendations for the use of coca as a stimulant, for digestive disorders, for treatment of cachexia, in withdrawal of morphine and alcohol addiction, for asthma, as an aphrodisiac, and local uses.

The arrangement and tone of this paper combined to create a tremendous general interest in coca and cocaine. Freud succeeded in his attempt to popularize coca, in part, because he recognized the coca controversy and, for the first time, exposed it under a single comprehensive title. Freud's pursuasive paper offered explanations for coca's previous failures and summarized its successes. Although portions of *Über Coca* are erroneous, it is universally recognized as the first important step in a series of closely related events that finally brought the coca controversy to a dramatic conclusion.

In June of 1884, Freud hastily finished *Über Coca* in order to prepare for his long-awaited trip north to visit Martha Bernays, whom he had not seen in two years. He sent his manuscript off to *Zentralblatt für die ges Therapie,* where it was immediately accepted and in print a month later. As soon as the article was published, Freud interrupted his experimental work with Koller and Konigstein and left for a holiday with his fiancée. Just before his departure, he wrote to Martha Bernays and at once revealed his casual personal use of cocaine and the reason he decided to interrupt his potentially lucrative research with the drug. He said, "I won't be tired [on the trip to Hamburg] becuase I shall be traveling under the influence of coca in order to curb my terrible impatience. . . . I want to forget everything connected with Vienna in your presence."[25]

While Freud was away in the north, his friend Carl Koller picked up a copy of *Über Coca* and unwittingly began the final preparation for a discovery that would make him famous.

Koller, an opthalmologist, had for some time been interested in the development of a workable local anesthetic suitable for eye surgery. The single local-anesthetic method available at the time was the Richardson ether spray, but this was only practical for short-term application in subcutaneous skin abscesses and similar operations. Medical science was desperately in need of a locally numbing agent to simplify operations where general anesthesia was unsatisfactory; this was especially true in eye surgery. Sometime before the dynamometer tests with Freud took place, Koller "began to experiment in local anesthesia of the eye with a view towards surgery—performing a great many experiments on animals. . . . [He] tried chloral, bromide and

**The discovery of local anesthesia.** ***(Courtesy of the artist, Darrell Orwig)***

morphine and other substances, but without success and gave up these experiments for the time being."[26] Koller later reflected on the importance of this early work, saying, "Although these experiments had been unsuccessful they had the good effect that my mind was prepared to grasp the opportunity whenever I should encounter a real anesthetic."[27]

The opportunity that Koller hoped for arrived one day shortly after the publication of Freud's landmark essay. The particulars of this great event are best described by the only two persons who were directly involved, Koller himself and Dr. Gaertner, the assistant in the laboratory of Professor Stricker. "Upon one occasion," wrote Koller, "another colleague of mine, Dr. Engle, partook of some cocaine with me from the point of a penknife and remarked 'How that numbs the tongue.' I said, 'Yes, that has been noticed by everyone that has eaten it.' And in the moment it flashed upon me that I was carrying in my pocket the local anesthetic for which I had searched some years earlier."[28] At this point, Gaertner continues the story:

> Dr. Carl Koller ran into Professor Stricker's laboratory, pulled out a small bottle containing a trace of white powder from his pocket and addressed me as Stricker's assistant, with a speech, the essence of which was: "I hope, indeed expect, that this powder will make the eye insensitive to pain." "We'll find out about that right away," I replied. We trickled the cocaine solution under the upraised lids of each other's eyes. Then we put a mirror before us, took a pin in hand and tried to touch the cornea with its head. Almost simultaneously we could joyously assure ourselves "I can't feel a thing." We could make a dent in the cornea without the slightest awareness of the touch, let alone any unpleasant sensations or reactions. With that the discovery of local anesthesia was complete. I rejoice that I was the first to congratulate Dr. Koller as a benefactor of mankind.[29]

After this historic event took place in Stricker's laboratory, Koller returned to his room at the hospital and excitedly began to organize a paper that would make public his recent discovery. He realized that in about an hour's time he had transformed cocaine from a little-known laboratory curiosity to the world's first completely effective local anesthetic, and he was anxious to publish this fact at the earliest possible opportunity. To this end, Koller wrote a brief preliminary communication that was completed in time for the Heidelberg Ophthalmological Society's meeting on September 15, 1884. Because he was so poor, the discoverer of local anesthesia was obliged to forgo the 300-mile trip to the conference and had to have Dr. Brettauer of Trieste read his paper before the society.

As mentioned, Koller's initial publication at Heidelberg was intended to be a brief announcement of his discovery; accordingly, it was very short and to the point. At the end of his carefully worded two-page statement, Koller said, "As I performed these experiments only during the past two weeks, I shall have to take up in a later publication the work which has previously been done on this subject."[30] Koller's chance to present a more complete account of his work came about one month later when he was asked to deliver a lecture on cocaine and anesthesia before the Viennese Medical Association.

After Koller read an expanded version of his Heidelberg paper, a most inappropriate event took place. Dr. Leopold Konigstein, the man whom Freud had interested in cocaine and eye diseases, addressed the medical association and announced that it was *his* experiments that had shown cocaine's usefulness in the treatment of certain eye afflictions and that it was also a perfect local anesthetic. His failure to mention Koller's similar findings in two previous papers was a conspicuous attempt to usurp Koller's priority in this matter and almost led to a nasty confrontation. Luckily, Freud had just returned from his holiday at this point. When he learned of the situation that had developed between his two friends, he quickly interceded on behalf of both men and began to negotiate a peaceful settlement. He wrote to Koller:

> Dear Friend:
>
> I am aghast at the fact that in K's published paper there is no mention of your name; and I don't know how to explain it in view of my knowledge of him in other respects; but I hope you will postpone taking any steps until I have talked to him, and that you will, after that, create a situation in which he can retract.
>
> With kind regards,
>
> Dr. Sigm. Freud[31]

Freud did indeed speak to Konigstein and urged him to correct his sin of omission. When a paragraph of recognition was finally added to Konigstein's paper, Koller graciously accepted it despite the fact that when Konigstein first heard that Koller declared cocaine to be a perfect local anesthetic for eye operations, the senior ophthalmologist insisted that Koller was wrong, a position he quickly abandoned and, obviously, later reversed.

While Konigstein's reference to Koller appeared to settle the matter of precession in the discovery of local anesthesia, it really only succeeded in setting the stage for a nearly identical battle over appreciation that featured Koller this time as the offender rather than the offended. Just as Freud was angered by Konigstein's failure to acknowledge Koller's previous work, so Professor Rossbach of Würzberg was now en-

**Dr. Carl Koller.** ***(Courtesy of Mrs. Hortense Koller Becker)***

raged over Koller's supposed neglect of the earlier work carried out by von Anrep.

Koller torpedoed this accusation in a pithy letter to the editor of *Berlin Klin. Wochenschrift,* the journal in which Rossbach aired his grievance. It is probably more than a coincidence that Dr. Leopold Konigstein was the editor of *Berlin Klin. Wochenschrift* at the time, and one can only wonder what went through his mind as he read Koller's letter.

The communication, dated December 17, 1884, contained a concise three-part rebuttal that completely silenced Rossbach and must have had a paralyzing effect on Konigstein as well. To begin with, Koller pointed out that Rossbach had previously based his criticisms on a review article he had read written about the Heidelberg report and the reason that he missed the reference Koller made to von Anrep's paper was because it simply was not mentioned in the review. Koller also said that, if somehow Rossbach did read the original Heidelberg report, then "I . . . regret very much that Herr Rossbach did not look at the wording."[32]

The second point Koller made in his letter was this: "There can be no question of von Anrep's priority in regards to the anesthetic effect of cocaine on the mucous membrane, since this was already known to the first researcher about cocaine in Europe . . . as well as to all those who followed."[33] In other words, Koller realized that he was not the first to make the discovery of cocaine's anesthetic abilities, but neither was von Anrep! By stating this, Koller set up his third point, which was, in effect, a definition of precisely what he considered his discovery to be. He said, "I have never taken credit in regards to the discovery of this useful physiological characteristic of cocaine, although its effect on the cornea was never before attempted. I have only made that step, as Professor Rossbach rightly remarks, to turn well known or easily deduced effects of cocaine into use in practical medicine, especially in the field of ophthalmology."[34]

What Koller did not mention, of course, was that the "step" he referred to was one of the most important in the history of medicine and that he took that step before anyone else and in the face of sizable odds.

Unhappily, Koller was still the object of some severe criticism over seventy years after he made his discovery. Freud's biographer, Ernest Jones, believes that a bibliographical error in Koller's second paper, giving the publication date of *Über Coca* as August 1884 instead of July 1884, was done purposefully to give the impression that Koller's discovery was simultaneous with Freud's publication and not just after it. One need only turn to the paper in which the error was committed to learn that in the very same text Koller made reference to Freud and *Über Coca,* saying, "Cocaine was brought to the foreground of discussion for us Viennese by the thorough compilation and interesting therapeutic paper of my colleague at the General Hospital, Dr. Sigmund Freud."[35] This would hardly be the correct move to make for a man who was intentionally trying to create the impression that the publication of *Über Coca* and the discovery of local anesthesia were coincidental independent inventions. Even considering Koller's error, the August date he mistakenly gave for the publication of *Über Coca* would still not make that event synchronous with the September date of the discovery of local anesthesia. On top of all this, Dr. Goran Liljestrand of the Karolinska Institute points out that Koller's error should not be considered all that unusual because Konigstein made the exact same mistake in a paper he published only two days after Koller's.[36]

# 9

# The First Cocaine Disasters in Europe and America: A Tale of Two Addictions

In spite of the criticisms that surrounded this event, the news of Koller's discovery spread around the world in an amazingly short period of time. Articles announcing cocaine's wonderful abilities appeared in leading medical journals, as well as in popular magazines and the daily press. Koller himself was besieged with hundreds of letters from all over the world asking for help and advice.* Doctors on three continents began to work on additional uses for the drug and the pharmaceutical companies reacted to cocaine's popularity by tripling the price and initiating expeditions to South America to learn more about the parent plant.

The firm of Parke Davis and Company in Detroit, Michigan, made arrangements for their consulting field botanist, Dr. H. H. Rusby, to travel to Bolivia to "investigate the origin, production and native uses of the drug, as well as to establish connections for obtaining supplies of it."[1] They found that no real quantity of coca existed in the United States except for some small experimental lots that had been kept around for a long time. Parke Davis and Company wanted to become the first American firm to secure a large amount of coca in anticipation of the financial spin-off that was sure to follow Koller's discovery.

*A letter received from an American Cavalry officer even asked if Koller could travel to the United States to examine his horse.

Rusby left New York on January 10, 1885, aboard the S.S. *Acapulco* and arrived in Lima approximately one month later. He immediately set off for the *yungas* and encountered his first *cocal* four leagues outside the city of Crucero. Rusby purchased the entire crop of this plantation and hired his own pickers to gather and package the leaves when they were ready and ship them back to George Davis in Detroit. In the meantime, Rusby continued on to Chile to collect samples of the drug *chekan* and investigate some new sources of cinchona.

By the time Rusby returned to his rented *cocal,* he learned that the crop had already been picked and sent off to the United States, apparently completing the financial coup engineered by Parke Davis. When Rusby reached La Paz and began to make final preparations for his triumphant return to Detroit, he was informed that a Colombian revolution had caused an embargo on transportation and that all the Parke Davis leaves were still lying on the Pacific side of the isthmus. To make matters worse, George Davis cabled Rusby the next day to inform him that the company's desire to strike first had been shattered because other pharmaceutical companies in the United States had already received large shipments of coca that had come around the Horn. He informed Rusby that the price of cocaine had fallen from 75¢ to 3¢ a grain.

This depressing news inspired Rusby to stay in

**H. H. Rusby.** ***(From* Jungle Memories *by H. H. Rusby, 1937; used with permission of McGraw-Hill Book Co.)***

South America for a while to try to salvage something from his mission. In an effort to do this, he came up with the idea of manufacturing a raw alkaloid in the coca-growing regions and then shipping this less volatile substance back to the United States for refinement. For his experiments, Rusby set up a homemade still in a vacant room of the hotel where he was staying and began trying different methods of producing crude cocaine. Once, while he was attempting to distill off the alcohol from a five-gallon lot of extract, he caused a huge explosion that sent columns of burning gas up to the ceiling of his apartment. Rusby grabbed the red-hot still and threw it out over the veranda sending a stream of liquid fire down to the ground floor. This, in turn, set the veranda on fire, but fortunately the blaze was extinguished when the alcohol burned itself out. In his memoirs, Rusby reflected on the gravity of the situation, remarking that the hotel was situated in the middle of the city and had no water supply except that which was carried on the backs of men. He mentioned that if the building had caught fire, "nothing could have prevented a dreadful catastrophe."[2]

Although he almost destroyed downtown La Paz in the process, Rusby succeeded in making his manufacturing idea a prophetic one. In another six years, foreign-owned factories would be established all over Peru and Bolivia for the production of crude cocaine. As mentioned, this substance was far less volatile than the leaves, so it was easier and safer to send the product on long ocean voyages. Crude cocaine extract was lighter, more compact, and its manufacture in South America cut production costs through the exploitation of cheap labor. All these factors combined to make shipments of bulky, delicate coca leaves unnecessary, unprofitable and, for a time, almost unavailable.

While Rusby's new method seemed practical in every way, his economic shortcut was in reality a double-edged sword. Not only did it advance the accessibility of potentially dangerous cocaine, but at the same time it all but eliminated shipments of whole coca leaves, a substance whose value as a medicine was and still is in doubt because of lack of availability and experimentation. This unfortunate turn of events also pressed doctors and later some patent-medicine manufacturers to turn to cocaine in an effort to obtain the therapeutic effects they desired and expected from whole coca.

In all fairness to Rusby, he did warn that "the properties of cocaine, remarkable as they are, lie in an altogether different direction from that of coca," and he could not have known that his recommendation would drastically reduce shipments of whole coca leaves and encourage the same type of confusion that surfaced after Niemann first discovered the alkaloid and made it so exclusively popular.

Once more, laymen and scientists reasoned that if coca worked, then cocaine must work even better, and again the leaf was put aside in favor of its most dangerous component.

The dangers involved in cocaine use can be dramatically illustrated by the story of Dr. William Halsted, one of the first American physicians to receive and experiment with Parke Davis cocaine.

Halsted began practicing medicine in New York in 1870 and soon became famous for his surgical techniques and advanced research methods. In early 1885, shortly after the presentation of Koller's second paper, Halsted discovered that cocaine could be put into a hypodermic syringe and injected directly into a nerve to produce a state of local anesthesia in the entire area serviced by that nerve. In the process of developing and perfecting nerve-block anesthesia, however, Halsted and all his assistants be-

**William Halstead, America's first cocaine addict.** ***(Courtesy of Northwestern University, Archibald Church Medical Library Portrait Collection)***

came addicted to cocaine. Halsted published a few short papers on his findings in 1885, but never mentioned the fact that he had become America's first known cocaine addict. In 1886, he was unable to continue his writing or his medical practice, for by that time his need for cocaine had grown to the enormous amount of 2 grams a day.

Just after the publication of his last paper on cocaine anesthesia, Halsted mysteriously disappeared. He showed up months later in New York a confused and very sick individual. Later it became known that several of Halsted's trusted friends had chartered a sailboat and took him on an extended ocean voyage to help him kick the cocaine habit. Unfortunately, even several months at sea were not enough to cure the man who was once the captain of the football team at Yale and possibly the most gifted and popular young doctor in New York City. He still suffered from severe anxiety attacks and acute paranoia that combined to reduce his practice and personality to a shambles.

Later that same year, Halsted admitted himself to Butler Hospital in Providence, Rhode Island, and after what has been called an "epochal struggle," he declared himself cured and was able to return to his medical practice. He later went on to become one of the world's most respected and admired surgeons and was also asked to be one of the founding fathers of the prestigious Johns Hopkins Medical School. The tale of Halsted's addiction seems to have had a storybook ending, but this is not the case. The reality of his situation was brought to light eighty-three years after his release from Butler Hospital when the members of the board of trustees at Johns Hopkins opened a book that had been sealed with a silver lock. The book was a secret history of the Hopkins written by Sir William Osler, and it revealed for the first time that Halsted had indeed cured himself of his cocaine habit, but only at the expense of becoming a morphinist.

While Halsted was taking morphine to escape his cocaine habit, Freud's friend Fleischl was still taking cocaine in an attempt to eliminate his dependence on morphine. This paradoxical situation did not last long because by July 1885 von Fleischl's daily concaine consumption reached a full gram and he began to suffer from convulsions, insomnia, and a horrible cocaine-produced nerve symptom called formication, which gave him the sensation that insects were crawling under his skin. The man who was once thought to be the first person in Europe to be cured of morphine addiction by cocaine was now, sadly, the first person in Europe to acquire what could be called a cocaine addiction.

Shortly before this happened, Friedrich Erlenmeyer, the editor of *Centralblatt für Nervenheilunkde,* published a sharp criticism of Freud and his attempts to end morphine addiction with cocaine. Erlenmeyer was writing in reaction to a general alarm that appeared in Germany over earlier reports of cocaine's toxic effects. He called cocaine "the third scourge of mankind" and cautioned doctors against its use.

In response to Erlenmeyer's attack, Freud read a paper before the Psychiatric Society of Vienna called *On the General Effects of Cocaine.* In his rejoinder to Erlenmeyer, Freud stated: "I have no hesitation in recommending the administration of cocaine for such withdrawal cures in subcutaneous injections of 0.03 to 0.05 grams per dose, without fear of increasing the dose."[3] When Freud wrote this (March 1885), he was convinced that the drug was helping to cure his friend's addiction and he looked forward to Fleischl's complete recovery; just three months later, Freud was forced to write that Fleischl had been greatly harmed by the "frightful doses" and he gave him only six months more to live.

Even after Fleischl had increased his intake of cocaine to a dangerous level and began to show signs of having an unmistakable toxic psychosis, Freud still rejected the idea that these symptoms had anything to do with his cocaine use. While it is painful to learn of Freud's stubborn insistence in the face of his friend's imminent death, it is at the same time comforting to realize that Freud was not without some good reasons for his beliefs about cocaine and addiction.

First, it is important to remember that, as a coca scholar, he was well aware that both coca and cocaine were poorly understood substances with long histories of defamatory personal opinions tacked on to their use and effects. For this reason, Freud was suspicious of any criticisms of his "wonder drug" no matter how reasonable or obvious they appeared. In this respect it may be said that Freud was somewhat paranoid and, considering the way things turned out, this seems to have been the case. Nevertheless, there is a sizable collection of other facts that argue in favor of Freud and his convictions.

Freud had done a considerable amount of experimenting with cocaine, both on himself and on others such as Koller and Konigstein, whose judgment he respected and trusted. Throughout these experiments, there was never any evidence of addiction among the participants nor was there a record of any adverse reaction noted in any of the numerous tests. Freud had also been mailing cocaine to Martha Bernays and his sisters and he knew from their letters that the drug had no deleterious effect on them either. When Freud wrote *On the General Effects of Cocaine,* he was sure that he had never seen or heard of a single, well-documented case of cocaine addiction.

Freud was also favorably disposed to cocaine use because, thanks to that drug, his father had become one of the first to enjoy a painless glaucoma operation. Koller and Freud administered the cocaine anesthetic and Konigstein performed the operation. Freud proudly (and generously) noted that, on this occasion, the three people who had been responsible for the discovery of local anesthesia were all together at the same time.

Freud's idea about the uses of cocaine were given the support of several distinguished scholars, including Dr. Bauer, Fleischl's personal physician. Fleischl himself added a note to an article Freud wrote in December of 1884, describing his own good experiences with the drug. Leonard Corning, the man who has been called the father of spinal anesthesia, wrote that there was a morbid fear of cocaine spreading throughout the country and he urged physicians not to be prejudiced against "a most useful remedy."[4]

William Martindale, who later became the president of the Pharmaceutical Society of Great Britain, published a popular little book entitled *Coca, Cocaine and Its Salts.* In this book, Martindale recommends the medicinal use of coca and cocaine against a variety of illnesses including morphine and alcohol addiction; he even goes so far as to suggest that the English substitute coca in place of their daily tea. Martindale's remarks carried a significant amount of weight with them because he was also the author of the much-respected *Extra Pharmacopoeia,* a standard reference work that is still found on the shelves of pharmacies throughtout the United Kingdom.

Freud also gathered a tremendous amount of support from the famous William Alexander Hammond, surely one of the most colorful figures of his day. Hammond first gained public attention in 1863 when he was named surgeon general of the United States Army. The thirty-five-year-old Hammond was to suffer a court martial at the hands of Abraham Lincoln a year later, but he recovered from this setback and went on to establish a respectable hospital in Washington, D.C., for the treatment of nervous diseases. From his position in the hospital, Hammond earned an international reputation as a neurologist. In addition to his medical credentials, he was also famous for writing plays and novels as well as books on sleep, neurology and impotency.[5]

Hammond presented a paper at a meeting of the New York Neurological Society entitled *Coca: Its Preparations and Their Therapeutic Qualities, with Some Remarks on the So-Called Cocaine Habit.* He joked about cocaine addiction, saying: "On four different days, I gave myself an injection. And, gentlemen, I experienced none of the horrible effects, no disposition to acts of violence whatsoever; why, I didn't even want to commit a murder."[5] Hammond went on to explain how he increased the amount of cocaine he injected so he could explore the physiological parameters of the new drug. He mentioned the "peculiar thrill" that accompanied all his injections and described how this sensation enlarged to enormous proportions along with the amount of cocaine he was using. This description was occasioned by the results of his latest experiment, which featured the injection of eighteen grains of cocaine in four installments, spaced only five minutes apart. Hammond stated:

> In all the former experiments, although there was a great mental exaltation amounting at times almost to delirium, it was nevertheless distinctly under my control, and I am sure I could have obtained entire mastery over myself, and have acted after my normal manner. But in this instance, within five minutes after taking the last injection, I felt that my mind was passing beyond my control, and that I was becoming an irresponsible agent. I did not exactly feel in a rebellious mood, but I was in such a frame of mind as to be utterly regardless of any calamities that might be impending over me. I do not think I was in a particularly combative condition, but I was elated and possessed of a feeling as though exempt from the operation of deleterious influences. I do not know how long this state of mind continued, for I lost consciousness of all my acts within, I think, half an hour after finishing the administration of the dose. Probably, however, other moods supervened, for the next day when I came downstairs, three hours after my usual time, I found the floor of my library strewn with encyclopedias, dictionaries and other books of reference, and one or two chairs overturned.[7]

Hammond realized that his experience brought him close to consuming a fatal dose of cocaine and he mentioned this fact in his paper. What really impressed him, however, was the fact that he had taken a large amount of cocaine over several days' time and afterwards felt no need or compulsion to use it again. Hammond was surprised he didn't wake up addicted after only four days of use and a near-fatal dose. Like Freud, he was seemingly oblivious to cocaine's high potential for abuse and could not foresee the problems involved in making this concentrated product available in wholesale lots to the general public. A possible explanation for this is that Freud and Hammond were both repulsed by the drug whenever they took too much. Freud often spoke of his aversion to cocaine, "which was sufficient cause for curtailing its use."[8] And Hammond was known to have been wracked by severe migraines and constipation for days after his last experiment. Nevertheless, Hammond closed his paper by prescribing a wine glass of coca with each meal and announcing that cocaine had been recognized as the official remedy of the Hay Fever Association.

Possibly the greatest reason for Freud's belief that cocaine is nonaddictive lies in the fact that cocaine is not a physically addicting substance.[9] The type of addiction that Freud and Hammond were used to seeing (and were looking for in cocaine) involved a dramatic withdrawal crisis such as that experienced by opium or morphine users. These withdrawal symptoms included severe sweating, convulsions, diarrhea, hysteria, and vomiting, and they invariably began to appear approximately four to twelve hours after the last dose was taken.

This is simply not the case with cocaine. There are no such predictable consequences associated with the discontinuation of this drug.[10]* This fact understandably led Freud and others to believe that cocaine was nonaddictive. In addition to the absence of withdrawal symptoms, there also appeared to be no reduction of the effect of cocaine with infrequent low-dose administration or, in other words, no tolerance or need to increase the dose to achieve the desired effect as there was with other known addicting substances.

It should be mentioned that Fleischl found it necessary to increase the amount of cocaine he used, but this may be taken as evidence that tolerance sometimes does develop after the user's habit reaches a point when the amount of cocaine is unusually high.[11] This is perhaps one of the reasons that Freud argued because his cocaine use was infrequent, he had no need to escalate his dose, and because he did not increase the dose, he did not become addicted.[12] It has also been suggested that it takes an addictive personality to become a cocaine addict; others believe that there is a hereditary enzyme difference that makes some people addiction prone.[13]

Whatever the reasons for cocaine's unpredictable consequences, it is clear that Freud and many others were lulled into believing that the drug was harmless because it did not display the familiar addictive symptoms that are always associated with narcotic drugs. They failed to realize that cocaine is not a narcotic and that the dangers of using it rest not with the inevitable tolerance syndrome or the horrors of withdrawal, but with the drug's unbelievably high abuse potential and the onset of an ugly cocaine-produced toxic psychosis whenever it is abused.

The paranoiac fits and other calamities that accompany the misuse of cocaine have, as in the case of Dr. Halsted, driven some addicts to turn to morphine or heroin for relief. While this

*This should not be construed as invitation to cocaine use. Many cocaine abusers report drug-related physical problems that make heroin withdrawal symptoms seem pale by comparison.

seems to be like going from the frying pan to the fire, some of the frightening aspects of cocaine abuse make the move seem almost justifiable. In 1969, G. A. Deneau had the opportunity to observe rhesus monkeys that were allowed to administer large doses of cocaine to themselves. He reported that after a few weeks, "the monkeys showed an extremely rapid loss of muscle mass and grand mal convulsions became frequent. Behavior consistent with visual hallucinations (staring and grasping at the wall) and tactile hallucinations (continued scratching and biting the extremities, to the point of producing extreme wounds and even amputation of the digits) was constantly observed."[14] Fleischl, who, like the monkeys, was administering large doses of cocaine to himself, was in a terrible condition as a result of his drug use. Dr. Freud wrote that it was impossible to describe his condition because he had never seen anything like it. He sometimes sat up all night with Fleischl and later wrote that on these occasions "every note of the profoundest despair was sounded." Koller saw Fleischl in

**Ernst von Fleischl-Markow, the first cocaine addict in Europe.** ***(Courtesy of Mrs. Hortense Koller Becker)***

the height of his battle with cocaine and said that he was in the worst condition imaginable. He noted that Fleischl was badly shaken by paranoid hallucinations that were teeming with white snakes.[15]

By 1887 reports of what was being called "cocaine addiction" began to mount up and lend support to the earlier accusations of Erlynmeyer and the pharmacologist Lewis Lewin. Freud, who never lost an opportunity to remind scholars of his role in the rediscovery of cocaine, was now suffering the reproach of his peers as a result of his efforts. Even his friend and former supporter Heinrich Obersteiner published a paper describing chronic cocaine intoxication and associated delirium tremens that were similar to those experienced by alcoholics. These attacks thrust Freud into a defensive position and he found it necessary to prepare a paper to attempt to explain cocaine's irregular action and at the same time clarify his position on the use and misuse of the drug.

In July of 1887, Freud published *Craving for and Fear of Cocaine;* this would be the last paper he would ever write on the subject, although we know that he continued to use and even prescribe cocaine at least until 1895.[16] *Craving and Fear* begins with the revelation of a supposed flaw in Erlynmeyer's research. Freud informs us that the reason Erlynmeyer noticed adverse reactions in his patients was because he administered the drug in subcutaneous injections. Freud then goes on to point out that had Erlynmeyer followed his recommendation of using only oral doses, he would not have seen any transient toxic effects.

Not only was Freud wrong in believing that oral applications of cocaine were completely harmless, but worse, he seems to forget that he continually encouraged Fleischl to inject himself with cocaine. He even left a lasting testimony to this mistake in March of 1885 when he "recommended the administration of cocaine . . . in subcutaneous injection" in *On the General Effects of Cocaine.*

In the next paragraph of his last paper on cocaine, Freud seems to retreat from his pro-coca position a bit when he admits that the alkaloid has no value for morphine addicts. He explained how addicts eventually substitute cocaine for morphine and make the former the more dangerous of the two. He states the case perfectly by saying, "Instead of a slow marasmus, we have rapid physical and moral deterioration, hallucinatory states of agitation similar to delirium tremens, a chronic persecution mania, characterized in my experience by the hallucination of small animals moving in the skin and cocaine addiction in place of morphine addiction."[17] Freud then ruined his accurate description of chronic cocaine abuse by declaring that these symptoms could only be found in cocaine users who were also morphine addicts. By doing this, Freud hoped to convince his readers that cocaine had claimed no victim of its own and that it had to be used by a susceptible, already addicted individual before it could do any harm. In this belief Freud was, of course, wrong again.

In what appears to be still another attempt to compromise his position on the substance that was dragging his reputation through the mud, Freud conceded that there did seem to be some clear-cut cases of acute cocaine poisoning (a condition characterized by stupor, dizziness, anorexia, and increased pulse rate). He suspected that the reason some users suffered this poisoning while others did not was the result of "individual variations in excitability and the variation of the condition of the vasomotor nerves on which cocaine acts."[18] Cocaine's unreliability truly bothered Freud and he took this opportunity to suggest that subcutaneous injections of cocaine be abandoned for the treatment of internal and nervous disorders. While this was a good idea, the problem with Freud's use of it was that he again mistakenly implied that the hypodermic needle was the real villain, distracting the reader from the dangers inherent in any heavy cocaine use no matter what the mode of administration.[19]

Freud had clearly backed off from his former position on cocaine. Later, he suppressed all reprints of and references to his paper advocating subcutaneous injections. In *Craving and Fear* he mentions that he had produced only two previous papers on the subject, *Über Coca* and *Contribution to the Knowledge of the Effect of Cocaine;* there was no mention of *On the General Effects of Cocaine.* He also failed to include his 1885 paper in a list of publications he assembled ten years later when he was applying for the position of full professor.[20] No copy of *On the General Effects of Cocaine* was found in Freud's library of papers and, in retrospect, it is most conspicuous by its absence. It has been suggested by several Freud scholars that these omissions were unconscious dishonesties and that they represent evidence of Freud's wish to repress this unfortunate episode in his life. Curiously, it was Freud himself who first recognized this subconscious trait and labeled it for science. By his own definition Freud had clearly committed a "parapraxis."

Freud's guilt and associated bad feelings over his recommendations of subcutaneous injections expressed itself only once, four years after Fleischl's death. In *The Dream of Irma's Injection* he said, "I had been the first to recommend the use of cocaine in 1885 and this recommendation had brought serious reproaches down on me. The misuse of that drug had hastened the death of a good friend of mine."[21]

By 1890, Freud achieved a goal he had always desired; he had become famous in medical circles all over Europe and America. Every doctor who had come to appreciate the dangers of heavy cocaine use learned to associate Freud's name with unreliability and recklessness. For his work with cocaine, Freud won the dubious honor of being the first scientist to be blamed for a major drug disaster.[22]

# 10

# Belle Epoque: Cocaine Adaptations in the 1890s

Although Freud suffered some serious professional rebuke because of his cocaine papers, there can be little doubt that his pioneer research inspired the medical community to examine a forgotten and poorly understood drug that badly needed examination. The publication of *Über Coca* inaugurated a decade of meaningful scientific investigations that produced much of what we know today about coca and its most famous alkaloid. In that ten-year period, coca was rescured from obscurity and vindicated against the long-held suspicion that it was an inert and possibly dangerous drug. Cocaine was experimented with seriously for the first time using human beings as test subjects and, unlike coca, was quickly proven to be a very unpredictable and sometimes toxic substance. Thanks in part to Dr. Freud's efforts, physicians everywhere were alive to the great harm that can come as a by-product of the imprudent use of cocaine. By 1890, reports of cocaine addiction were common in the medical literature and doctors were no longer confused about its ability to ruin the lives of those who were not careful.

Armed with this information, it is surprising to learn that the use of cocaine was about to come into vogue and seize the world as no drug plant since tobacco.

One would suspect that a substance with the proven abuse potential of cocaine would be considered undesirable by anyone who valued his health, or at least kept out of the hands of the insensitive and possibly still ignorant masses. Ironically, on both accounts, the exact opposite situation developed. Health-conscious individuals of the nineteenth century welcomed cocaine as an addition to their favorite over-the-counter medicines and, at the same time, cocaine was the object of an ambitious advertising and distribution campaign that studiously avoided any references to the drug's possible consequences.

Timing was possibly the key factor involved in this unpredictable burst of cocaine popularity. Between the time Niemann first isolated the drug and the publications of Freud's last cocaine paper, Europe and America modernized and embraced a social attitude reflective of the material, urban, and intellectual world that had suddenly been created. Unfortunately, this attitude reached a crescendo early in the 1890s, coincidental with the news of cocaine's dangerous potential.

The last decade of the nineteenth century has often been referred to as the "Gay Nineties," and for good reason. On October 6, 1889, the Moulin Rouge opened its doors in Montmartre and symbolically launched a decade of gaiety, elegant living, and convivial eccentric behavior. Creative people from all fields gathered together in great salons and advanced the tenets of

**An art nouveau coca maiden.** ***(Courtesy of the Fitz Hugh Ludlow Library)***

hedonism. It was the time of Oscar Wilde and Paul Gauguin, Jules Verne, and Toulouse-Lautrec.

Part of the social attitude of the nineties, often dubbed *fin de siècle,* or end of the century, concerned itself with the idea that the end of the century was arriving simultaneously with the end or death of traditional ideas and modes of behavior. There was a need for deliberate eccentricity and a desire to participate in the bizarre. The social climate dictated a departure from the ordinary and it produced an era of decadence, mysticism, and unusual behavior of all types. The nineties were also a time of intense creativity and the list of gifted contributors to the *fin de siècle* world of Western Europe would fill many pages. For some, this time period is fondly remembered as *la belle époque,* and nearly everyone has come to associate it (at least in part) with the sensual and the exotic.

The news that cocaine was a potentially dangerous drug could not have arrived in Europe at a worse time. In addition to being a famous anesthetic, the drug was also becoming well known for its ability to provide a uniquely pleasurable physical and mental experience, and this latter quality made it an irresistible companion for *la belle époque.* Cocaine euphoria would not be denied by the pleasure-seeking and mind-expanding personalities of the 1890s; they highlighted the drug's power to please and eagerly devoured it at the expense of all the tragic episodes that demonstrated the folly of zealous consumption. The pronouncement that cocaine use was a potentially hazardous enterprise fell on deaf ears; it might be said that the news was drowned out in the rush of enthusiasm over the drug's sensational effects coupled with the desire to achieve, as the poet Rimbaud (1854–91) said, "the systematic derangement of all the senses."

Free to develop in an almost unchallenged atmosphere, the popular use of cocaine quickly developed into a widespread fad. References to the drug spilled into the popular literature of the day and the shilling shocker as well as the dime novelette capitalized heavily on cocaine's recent popularity. As might be expected, more than a few of these stories were invented and frantically penned out by cocaine-influenced authors who had fired up their creative abilities with liberal amounts of that fascinating stimulant.

Recently it has been suggested that Robert Louis Stevenson was under the influence of cocaine while he was preparing *The Strange Case of Dr. Jekyll and Mr. Hyde.*[1] As everyone knows by this time, Stevenson created a horror classic when he lifted the curtain on the life of Dr. Harry Jekyll—a respectable London physician who was periodically transformed into a demonic madman by a powerful new drug that was

**Arthur Conan Doyle.** ***(Courtesy of Rev. John Lamond, D.D.)***

**Robert Louis Stevenson.** ***(Courtesy of Sir James Barrie, O.M.)***

**Actor William Gillette shoots cocaine into his wrist while playing Sherlock Holmes on the New York stage in 1901.** ***(Courtesy of the Fitz Hugh Ludlow Memorial Library)***

described as "a powder." Even more interesting is the fact that the seriously tuberculant Stevenson wrote and rewrote this story in an energetic six-day frenzy; his wife later remarked, "That an invalid in my husband's condition of health should have been able to perform the manual labor alone of putting sixty thousand words on paper in six days seems incredible."[2] The evidence suggests that Stevenson may have turned to cocaine in an attempt to relieve his failing respiratory condition and produced *Dr. Jekyll and Mr. Hyde* in the process. While the drug certainly did nothing to arrest his tuberculosis, it may have had the effect of providing Stevenson with the inspiration for a great story and the energy to see it through to a very speedy conclusion.

A once-celebrated London periodical called *The Strand* featured a number of stories that involved cocaine somewhere in the plot. A physician named Arthur Conan Doyle penned at least six of these for *The Strand* and even produced three in one year. Because he returned to the subject so often, we may assume that the use of cocaine was a popular if not fascinating topic for the majority of his readers; surely by this time many of them had actually used the drug in one form or another or else knew of plenty of people who did.

Dr. Arthur Conan Doyle's contributions to *The Strand* enjoyed a tremendous amount of attention because he created an exciting and unforgettable character in the person of Sherlock Holmes—the world's first consulting detective. Doyle made Holmes the epitome of what every nineteenth-century man wanted to be; he was brilliantly deductive, benignly eccentric, and patently irrepressible. To enhance these traits, Doyle often portrayed the indefatigable Holmes with a syringe in hand, injecting himself with a solution of cocaine hydrochlorate.

In *The Sign of Four,* Holmes was asked why he took cocaine. The great detective replied: "My mind rebels at stagnation. Give me problems, give me work, give me the most abstruse cryptogram or the most intricate analysis and I am in my proper atmosphere. I can dispense with artificial stimulants. But I abhor the dull routine of existence."[3]

Like Carl Koller, Arthur Conan Doyle was an eye doctor and as such must have been especially aware of cocaine's abilities and latent dangers. It

**Angelo Mariani's coca salon.** ***(Courtesy of the Fitz Hugh Ludlow Memorial Library)***

is interesting then that Doyle would have Holmes's companion/physician, Dr. Watson, regularly object to his friend's cocaine use, but only in an offhand way that would leave plenty of room for some convincing last words from the story's real hero. In one instance, Watson casually asked Holmes, "Which is it today, morphine or cocaine?" Holmes identifies the substance in his hand as cocaine and asks the doctor if he would care to try some. "No, indeed!" Watson replies, "My constitution has not got over the Afghan campaign yet. I cannot afford to throw any extra strain on it." At this point Holmes admits, "Perhaps you are right, Watson. I suppose that its influence is physically a bad one. I find it, however, so transcendentally stimulating and clarifying to the mind that its secondary action is a matter of small amount."[4]

Holmes's feelings about cocaine were perfectly typical of the times in which he lived. His eccentric fear of the "dull routine of existence" and his cavalier willingness to assume the hazards of cocaine's "secondary action" (in spite of sound medical counsel) place him squarely in the nave of late nineteenth-century drug consciousness.

Cocaine's sudden popularity during the Gay Nineties also pumped new blood into the coca-product industry and made millionaires out of Angelo Mariani and a few of his imitators.

In the 1890s French philosopher Henri Bergson filled elegant salons with audiences who came to hear him discuss the importance of *élan vital,* or the life force that was so essential in man's evolution. To his eager listeners, Bergson's words were not only significant, but also somehow strangely familiar. It did not take a great imagination to make the connection between a philosophy that advocated a powerfully alert body and mind and the well-known advertising rhetoric of a tonic stimulant that promised to fortify and refresh overworked bodies and brains. Thirty years of consistent production

**The artist Mucha and his tribute to Vin Mariani.** ***(Courtesy of Fitz Hugh Ludlow Memorial Library)***

and ambitious advertising had finally paid off, making Angelo Mariani one of the great tycoons of the patent-medicine era.

With sales and production booming, Mariani began construction of a number of new buildings that were all devoted to coca. At Neuilly on the Seine, apart from his office on the Boulevard Haussmann, Mariani set up a laboratory complete with expansive conservatories and a great salon. The complex was a magnificent example of turn-of-the-century architecture, resplendent in glass, steel, and wrought iron. In terms of coca art, the interior of Mariani's salon could only be matched by the Emperor Pachacuti's *Coricancha.* There were art nouveau coca designs by Courboin, leather models of the plant by St. André, and coca nymph statuettes by Rivière. In his salon, Mariani walked on coca rugs and sat on chairs with carved coca-leaf backs. It has been said that he took great delight in experimenting with the thousands of plants that he had growing in his greenhouses and it is known that he made different specimens available at no charge to any botanical garden in the world that could take proper care of them. A contemporary of his claimed that "coca is the hobby of Mariani. It is his recreation, his relaxation and his constant source of pleasure."[5]

Mariani's wine was drunk, enjoyed, and even endorsed by the most famous people in the world. While other patent-medicine manufacturers were scratching about for credible testimonials, Mariani easily filled thirteen large octavo volumes with page after page of warm notes of appreciation from the intellectual, political, and even spiritual giants of his day. Each approving paragraph was accompanied by an etched portrait and sometimes, as in the case of the artist Mucha, even a small sketch that featured a mysterious coca maiden holding a bottle of Vin Mariani.

The crowned heads of Mariani's book included Albert I (Prince of Monaco), Alfonso XIII (King of Spain), George I (King of Greece), Mozaffer-et-Dine (Shah of Persia), Oscar II (King of Norway and Sweden), Peter I (King of Serbia), Nicholas II (Czar of Russia), the Prince and Princess of Wales, and even King Norodom of Cambodia. Some of the well-known authors represented were Jules Verne, H. G. Wells, Octave Mirbeau, Henrik Ibsen, Anatole France, Emile Fabré, Alexandre Dumas, Edmund Rostand, and Emile Zola. Two presidents of France wrote in with their pictures as well as President McKinley and representatives of former President Ulysses S. Grant. Other notables included

**Edison drank Mariani's wine and is remembered as being able to work long hours with little sleep.** ***(Courtesy of Fitz Hugh Ludlow Memorial Library)***

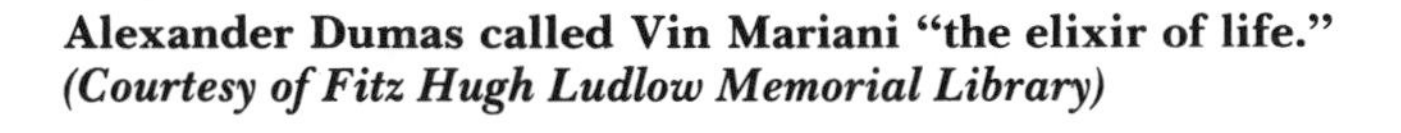

**Alexander Dumas called Vin Mariani "the elixir of life."**
***(Courtesy of Fitz Hugh Ludlow Memorial Library)***

Sarah Bernhardt, composers Jules Massenet and Charles Gounod, the sculptor François Rodin, Jean Lorrain, August Bartholdi, the creator of the Statue of Liberty, and, last but not least, Thomas Alva Edison, who was renowned for his abilities to work with little or no sleep.

Over 8,000 physicians gave their consent to Vin Mariani and the wine even won the ecclesiastical approval of two popes; Pope Pius X sent a nice letter to Paris and Pope Leo XIII went so far as to strike a gold medal for Mariani as a token of his material gratitude. In addition to this, Leo XIII added a note that assured Mariani that His Holiness had been supported in his ascetic retirement by a flask of coca wine which was "never empty."

When Mariani began production in 1863, his intention was to manufacture a medicinal tonic that would be purchased for its therapeutic abilities. By 1890, however, his wine had grown to become a fashionable drink that was noted for its recreational as well as medicinal capabilities. In the Gay Nineties it was no longer necessary to feel sick before partaking of a few glasses of coca wine; people who felt good were drinking it to feel even better. In this way, Vin Mariani moved out of the medicine cabinet and into the salons and dance halls of Paris, where it was relished as an active ingredient that seemed perfectly suited for debate as well as debauche.

The success of Mariani's tonic inspired a host of coca products to appear on the market. In the soft-drink category alone there was Cafe-Coca Compound, Doctor Don's Kola, Delicious Dopeless Koca Nola, Kola Ade, Kos-Kola, Inca Cola, Nichols Compound Kola Cordial, Kumforts Coke Extract, A. L. Pillsbury's Coke Extract, Vani Kola, and even Rococola.

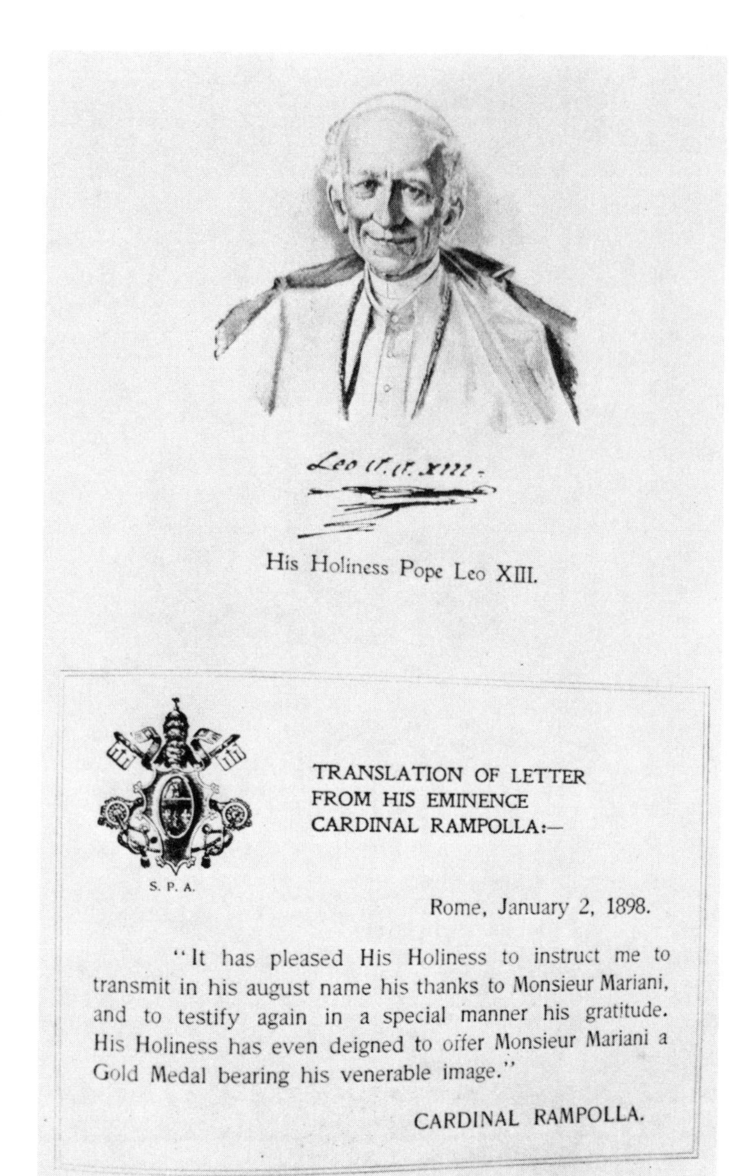


S. P. A.

TRANSLATION OF LETTER
FROM HIS EMINENCE
CARDINAL RAMPOLLA:—

Rome, January 2, 1898.

"It has pleased His Holiness to instruct me to transmit in his august name his thanks to Monsieur Mariani, and to testify again in a special manner his gratitude. His Holiness has even deigned to offer Monsieur Mariani a Gold Medal bearing his venerable image."

CARDINAL RAMPOLLA.

**Pope Leo XIII struck a medal for Mariani as a token of his material gratitude.** ***(Courtesy of Fitz Hugh Ludlow Memorial Library)***

THE MEDICAL RECORD.

**COCA BEEF TONIC,**

**COCA BEEF TONIC, with Citrate of Iron.**

**Highest Medals—Endorsed by Leading Physicians.**

*Having been made acquainted with the mode of preparation and the composition of the Coca Beef Tonic, I have ordered it for patients requiring tonic treatment from defective performance of the functions of assimilation; such patients derived marked and decided benefit from its administration, and I would recommend its use in cases where the system is suffering from impoverished blood and enfeebled nutrition. Scientific men are becoming more and more impressed with the necessity of supplying, by nutritive ingesta, the wear and tear of civilized life, and the Coca Beef Tonic is composed of materials well suited to fulfill the necessary requirements for which it has been prepared.*

J. M. CARNOCHAN, M D.

Coca Beef Tonic embodies the nutritive essence of carefully-selected choice beef, combined to an elixir of Coca (Erythroxylon), the two being in solution in a choice first-class quality of Sherry Wine. Coca Beef Tonic, with Citrate of Iron, in addition to the ingredients named, contains two grains Citrate of Iron to the tablespoonful.

Full formulæ on the bottles.

**A turn-of-the-century coca preparation.** ***(From*** **Medical Record*****)***

Mariani's greatest imitator, however, was an American calvary officer named John Styth Pemberton. Pemberton arrived in Atlanta, Georgia, in 1869 with a little over a dollar in his pocket and began to brew a couple of nostrums known as Triplix Liver Pills and Globe Flower Syrup. In 1885, Pemberton registered a trademark for an item he called French Wine Cola—Ideal Nerve and Tonic Stimulant. Some combination of competitive strain from Mariani and pressure from different prohibitionist groups convinced Pemberton to change the name of his drink and replace the wine with caffeine. One year later he was selling "Coca-Cola—an intellectual beverage and temperance drink."*

In 1891, Pemberton sold the entire rights to Coca-Cola to a pharmacist named Asa Griggs Candler for $2,000; in the same year, Candler acquired the exclusive bottling rights to something he called Botanic Blood Balm. For a while, the pharmacist's two new patent medicines were alternately brewed in the same iron pot, which was located in his backyard. Later Candler decided to concentrate all his energies on the production of Coca-Cola. Along with this decision came a significant shift in advertising that would rearrange the future course of one of the world's most sought-after beverages.

Students of the Coca-Cola story have pointed out that Candler began to emphasize the pleasure-giving rather than the therapeutic qualities of Coca-Cola shortly after he took over.[6] This purposeful shift in marketing philosophy neatly coincides with the beginning of the Gay Nineties and the realization that products such as Coca-Cola and Vin Mariani could be used to enhance and sometimes even produce a good time.

COCA-COLA
SYRUP AND EXTRACT.
For Soda Water and other Carbonated Beverages.

This "INTELLECTUAL BEVERAGE" and TEMPERANCE DRINK contains the valuable TONIC and NERVE STIMULANT properties of the Coca plant and Cola (or Kola) nuts, and makes not only a delicious, exhilarating, refreshing and invigorating Beverage, (dispensed from the soda water fountain or in other carbonated beverages), but a valuable Brain Tonic, and a cure for all nervous affections — SICK HEAD-ACHE, NEURALGIA, HYSTERIA, MELANCHOLY, &c.

The peculiar flavor of COCA-COLA delights every palate; it is dispensed from the soda fountain in same manner as any of the fruit syrups.

J. S. Pemberton;
Chemist,
Sole Proprietor, Atlanta, Ga.

**An early advertisement for Coca-Cola.** ***(Courtesy of Fitz Hugh Ludlow Memorial Library)***

*Pemberton continued to use coca leaves in the preparation of his drink, and in fact coca leaves are still a component in the secret Coca-Cola formula. It is known however, that since the passage of the Pure Food and Drug Act in 1906, great care has been taken by the company to remove the cocaine alkaloid from the leaves before they go into the making of the beverage.

# 11

# Patent Medicines and Drug Laws: From Green Leaf to White Grief

By 1895, coca's recreational reputation had spread all over the world; this phenomenon was caused in part by advertising and in part by accident. It is not known how many people began to use some coca product as a specific and unsuspectedly ended up with a lasting appreciation of their former medicine's capabilities as a sport.

This situation became aggravated when some unscrupulous or unwitting patent-medicine manufacturers began to jump on the coca bandwagon. It did not take long for these manufacturers to realize that the simple addition of cocaine hydrocholorate to their product was a quick and easy substitute for the relatively elaborate blending process adhered to by Mariani and, to a lesser extent, Asa Candler.

Cocaine hydrochlorate was added to health-care products not because it was an effective cure (as was always advertised), but because the right amount of cocaine in the preparation made the patient *feel* as if he were cured. This "right amount" of cocaine was determined solely by the manufacturer's desire to make sure his medicine had a powerful and positive effect, so consequently doses tended to run quite high.

Cocaine hydrochlorate was most often found in tuberculosis and hay-fever remedies that were designed to be sniffed, injected, or brushed directly onto a mucous membrane. These products and their modes of application are in direct contrast to the medicines that were made with the whole coca leaf or its fluid extract because the latter were recommended only for the relief of simple fatigue and other minor discomforts and were almost always packaged to be drunk.

These considerations should have been particularly noteworthy for the coca user, because the difference between drinking a coca cordial such as Vin Mariani and sniffing a dangerous cocaine-laced preparation such as Dr. Tucker's Asthma Specific, for instance, was a tremendously significant one. In the first place, Vin Mariani contained only .10 grains of cocaine per ounce while Tucker's Asthma Specific contained anywhere up to 7 full grains per ounce; later analysis by the *Journal of the American Medical Association* revealed that Dr. Tucker and others haphazardly prepared their medicines with widely varying amounts of cocaine and went on to suggest that it would be rare luck to find any consistency of strength in these products.[1]

Next, and possibly just as important, is the fact that when Mariani's wine was drunk, the small amount of cocaine it contained underwent a slow absorption process in the gastrointestinal tract, preventing the user from obtaining a high blood level of the alkaloid, even when large doses were ingested.[2] The fatal dose of cocaine after oral ingestion is reported to be about 1,200 mg. Dr. Tucker, on the other hand, recommended that the application of his powerful "remedy" be made directly to the mucous membrane in the

**A warning from *Hampton's Magazine*. *(From* Hamptons Magazine *24, no. 1, 1911)***

nose and even sold an atomizer along with his nostrum to facilitate this process. Whenever cocaine is applied in this fashion, it is absorbed very rapidly, and dangerously high blood levels of the drug are quickly achieved. The fatal dose of cocaine after injection, sniffing, or any other topical application to a mucous membrane can sometimes be as low as 20 mg![3]*

Shielded in part by the success of the harmless coca drinks, many dangerously powerful cocaine products such as Dr. Tucker's began to slip on the home-remedy market. Before long, it was possible to walk down to the corner pharmacy and purchase Agnew's Powder, Anglo-American Catarrh Powder, Cassebeers Coca Calisaya, Dr. Birney's Snuff, or any one of the Cole, Grey, or Crown Cures, to name just a few. All of them contained enough cocaine to make a dramatic difference in the way one felt, and all came complete with detailed instructions on dosage and routes of administration that were tantamount to invitations for addiction.

This was a particularly bad state of affairs because many people who would have normally balked at the idea of using cocaine by itself cheerfully embraced the same substance when it was packaged and served up as a medicine. Soon, more than a few of these individuals found themselves habituated rather than cured and continued to buy and use cocaine products

*While it is true that just this small amount of cocaine could (and has) killed people, it is obvious that most people can tolerate much higher doses before they actually drop dead. The problem with cocaine abuse is not that relatively small doses are sometimes fatal but rather that habitual abuse produces a living death of associated physical problems as well as sociopathic behavior.

**An advertisement for Dr. Tucker's dangerous Asthma Specific. (*From* Journal of the American Medical Association)**

to support this habit. Apart from these innocent victims, there were, of course, those who always thought of the "cocaine cures" as merely euphemisms for ways in which to get high. From the beginning, they purchased products such as Dr. Tucker's solely for its cocaine content and the euphoria it was sure to deliver.

At the turn of the century, the combined total of accidental and purposeful users of cocaine was very high. The list of those who were familiar with the drug expanded to include such diverse groups as sharpshooters, school children, preachers, lawyers, athletes, and actors. As more and more people began to use cocaine, the number of them who began to abuse it grew in proportion; before long, the indulgences and excesses of a percentage of this latter group attracted the attention of a nation about to succumb to a hostile and ambitious temperance crusade.

Cocaine, rather than tobacco or alcohol, became the main focus of the prohibitionists' attack. This reaction was prompted, in part, by a report of the Committee on the Acquisition of the Drug Habit issued in 1902. The committee revealed that, in the previous five-year period, the population of the United States had increased by ten percent while the importation of cocaine had increased by forty percent; and this latter statistic did not reflect the amount of domestic cocaine produced by Parke Davis and others. This disturbing piece of news was compounded by the results of a survey released that same year stating that only three to eight percent of the total amount of cocaine sold in major U.S. cities was used for legitimate medical needs.[4]

These figures must have been particularly infuriating to the ardent prohibitionist, for not only did they indicate that cocaine use was increasing at a rate at least four times faster than the population, but worse, they seemed to confirm the suspicion that this substance was little more than an object of capricious gratification.

The idea of expressing outrage over the non-medical use of stimulants and narcotics was not a new one. Throughout the nineties and before, there were regular protests from moral-minded churchmen, purists and, of course, the Women's Christian Temperance Union, but before 1900 no one paid much attention to them.

Times change, however, and under the pressure of a long series of related events, public opinion began to make a conservative shift that tended to favor the interest of these groups. To begin with, by 1900 the patent-medicine industry had just about stretched its credibility to the snapping point. People were not quite so willing to be duped by the industry's various assaults on their pocketbooks and intelligence. The public's growing anger seemed justified in response to the stimulus provided by the so-called healers of the day.

Early into the century, flamboyant medicine shows toured rural America touting cure-alls that were sometimes wholly inert and other times dangerously active. For those citizens who did not go in for this type of burlesque, there were numerous mail-order catalogs for drugs, door-to-door drug salesmen, and established quack doctors located in every city who guaranteed to satisfy even the most whimsical prescription need.

Profit motive and health care are often a bad mix. In the absence of any kind of regulation, the two became a monster. In 1906, there was seemingly no end to the schemes and gimmicks of the freewheeling bunco artists who plagued the patent medicine industry. Their work produced confusion, toxicity, and addiction. It promoted a fear of drugs based on reinforced ignorance or lies and, in addition, prompted a growing national insecurity about the resolution of an obvious problem.

The fear of adulterated products was intensified by the publication of Upton Sinclair's *The Jungle* and brought to the boiling point by the investigative reporting of Samuel Hopkins Adams. Adams, who published a series of ten articles for *Colliers* magazine entitled *The Great American Fraud,* alerted the public to the hazards of using secret nostrums and accepting quack medicine in general. His articles in *Colliers* are also considered instrumental in the passage of the Pure Food and Drug Act in 1906, the legislation that marked the first congressional attempt to regulate the consumption of cocaine in the United States.

The heart of the Pure Food and Drug Act was its power to prohibit

> the introduction into any state or territory or the District of Columbia, from any other state or territory or the District of Columbia, or from any foreign country or shipment to any foreign country, any article of food or drug which is adulterated or misbranded under the meaning of this act.

Under the terms of the Pure Food and Drug Act, a food item or a drug was misbranded if it contained, but did not list on the label, alcohol, morphine, heroin, cocaine, or any derivatives or preparations of these substances. A confectionary, such as Coca-Cola, for instance, was considered adulterated if it contained *any* narcotic drug. The act also demanded that importers of any food item or drug product that was misbranded or adulterated or considered dangerous to people sign a statement swearing that the product was not intended to be used in a manner dangerous to health.

The first section of the act also made it illegal to manufacture adulterated or misbranded food or drugs, but only in the District of Columbia or within the bounderies of any U.S. territory. Congress agreed that it lacked the constitutional power to prohibit any of the states from manufacturing these items and so they endeavored to control the situation as best they could by regulating the interstate transportation of such goods.

The passage of the Pure Food and Drug Act can be viewed as a major advance in the field of public health. In essence, the 1906 law ended the practice of slipping secret additions into healthcare products by requiring that quantitative and qualitative labels be affixed to all medicines containing alcohol, morphine, opium, or cocaine. Such labels allowed the more enlightened members of the community to see at a glance that they were not being cured at all by these preparations and that, in fact, they were only being made drunk and possibly even sicker by the inclusion of some mind-altering drug.

While there can be little doubt of the Pure Food and Drug Act's impact on the popularity and marketing of certain preparations, the law's overall effectiveness in regulating the consumption and manufacture of dangerous drugs was questionable. For instance, it was still quite legal in most places to manufacture and use preparations containing drugs as long as the contents were clearly marked on the label, and provided that the product was not earmarked for interstate shipment. Those who wished to bring such products into the United States needed only to affix their signature to a government form saying that they had no intention of injuring the health of any U.S. citizen. Those who were caught disobeying the provisions of the new law were subject to modest fines ($200 maximum) and only repeated offenders faced the possibility of jail sentences.

It soon became clear to many people that the Pure Food and Drug Act was a nice thing to have around, but it was simply not adequate to deal with the nation's growing drug problem. As a result, several states enacted laws of their own to try to deal more effectively with the issue; but even then, the use of cocaine and other drugs continued to grow.

New York State, for example, passed the Smith Anti-Cocaine Bill in 1907, but judging from a *New York Times* article dated approximately one year later, the popularity of the drug in New York City seemed barely affected. The New York City Police Department reported that cocaine was easily obtainable all along Seventh Avenue, between Twenty-eighth and Twenty-third Streets, and that "snuff parties" or just "blowing the Birneys"* was a frequent and informal happening in other parts of the city as well.

Inspector H. H. Mastison complained that the penalties under the Smith Act were not strict enough and cited several of his arrest cases as

*Most likely a term referring to the social consumption of Dr. Birney's Catarrh Powder, in its day a popular tuberculosis cure.

**The home of "the coke king of Chicago" ca. 1910.** ***(From* Hamptons Magazine *24, no. 1, 1911)***

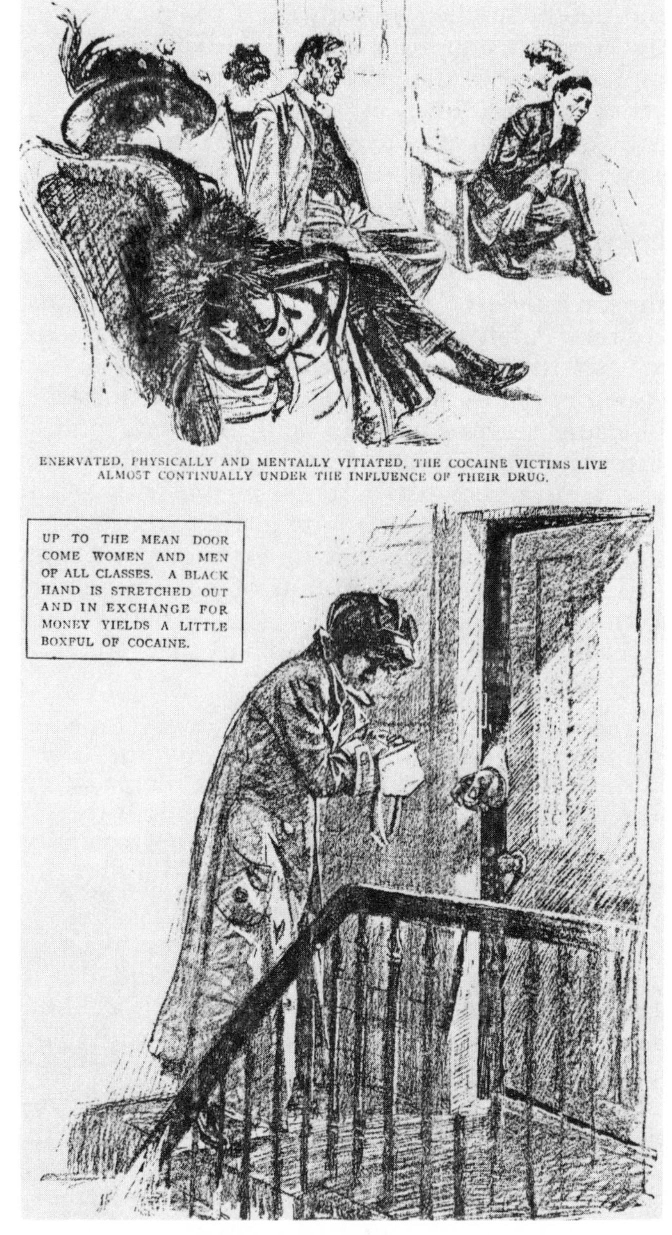


**Turn-of-the-century temperance propaganda.** ***(From* Hamptons Magazine *24, no. 1, 1911)***

evidence. He noted that Richard Flood, also known as Pork Chops, was one of the most notorious cocaine dealers in the city and his arrest on those charges resulted in only a six-month jail sentence. Margaret Lee, also known as the Irish Queen, received thirty days, and only recently Frank Smith, a cripple, got the full sentence of one year in jail. Monetary fines leveled under the Smith Act were sometimes as low as five dollars and, in any case, were relatively minuscule compared with the grossly inflated price of the drug since prohibition.

In the absence of an effective and comprehensive drug law, the concern over misuse of cocaine degenerated into an almost unchecked fear of the drug itself. In an attempt to produce a workable cure, temperance advocates took the situation into their own hands and launched a no-holds-barred attack on cocaine that advertised only its worst possible aspects. The philosophy seemed to be aimed at turning the tables on the drug's reputation by attempting to change its image from one of having a good time or possibly even getting well into a causal association with unhealthy, deviant, and even criminal behavior.

Pressured by a wave of emotionally charged reports, an anxious nation embraced many strange beliefs about cocaine. In an effort to bring the situation under control, some of these unfounded assumptions about coca and cocaine were just accepted as fact and then seized on as good reasons for the enactment of nothing less than maximum control measures.

Many of the current misconceptions about coca and cocaine have their roots in this early twentieth-century folklore. One of the most common mistakes is the belief that any indulgence in cocaine, no matter how small, will result in a horrible addiction. Consider these two excerpts from the *Medical Record* of February 1914: "There is no such thing as an occasional or moderate cocaine user. The line is very sharply drawn between the total abstainer and the fiend, without any intermediate condition."[5] And a few paragraphs later: "Once the Negro has reached the stage of becoming a dope taker (he means cocaine) . . . a very few experimental sniffs of the drug will make him a habitue."[6]

These extravagant exaggerations are echoed in another article published the same year in the journal *Military Surgeon* which states that "an individual succumbs promptly, and after several trials does not trouble to resist at all, another entry in Satan's ledger."[7]

The unnecessary fear generated by this type of hyperbole was surpassed only by an additional charge, formed then and still with us today, that says the use of cocaine causes violent acts and/or criminal behavior. In 1914, the *Literary Digest* warned that after sniffing cocaine, "the most daring crimes [are] committed."[8] In 1911, *Hampton's Magazine* ran a feature article on cocaine that typified the sort of sensationalism dredged up to convince the reading public that cocaine and crime were inseparable. The article discussed the case of Annie C. Meyers, a once-successful Chicago businesswoman who had turned into a cocaine fiend. After stressing the notion that she was at one time "a well-balanced Christian woman," Meyers proceeded to outline her downfall and emphasized the ways in which she believed cocaine was responsible. She said: "As the first effects of the drug are kleptomania, I was constantly in trouble." The article then mentioned that she had been arrested in Cleveland, St. Louis, Indianapolis, and Chicago for stealing things from department stores. Next, Meyers related a macabre tale of vicious self-mutilation that was also "caused" by her cocaine use: "I deliberately took a pair of shears and pried loose a tooth that was filled with gold. I then extracted the tooth, smashed it up, and taking the gold went to the nearest pawnshop (the blood streaming down my face and drenching my clothes) where I sold it for 80¢."[9]

Annie C. Meyers's depraved story is interrupted at this point for some comments from the

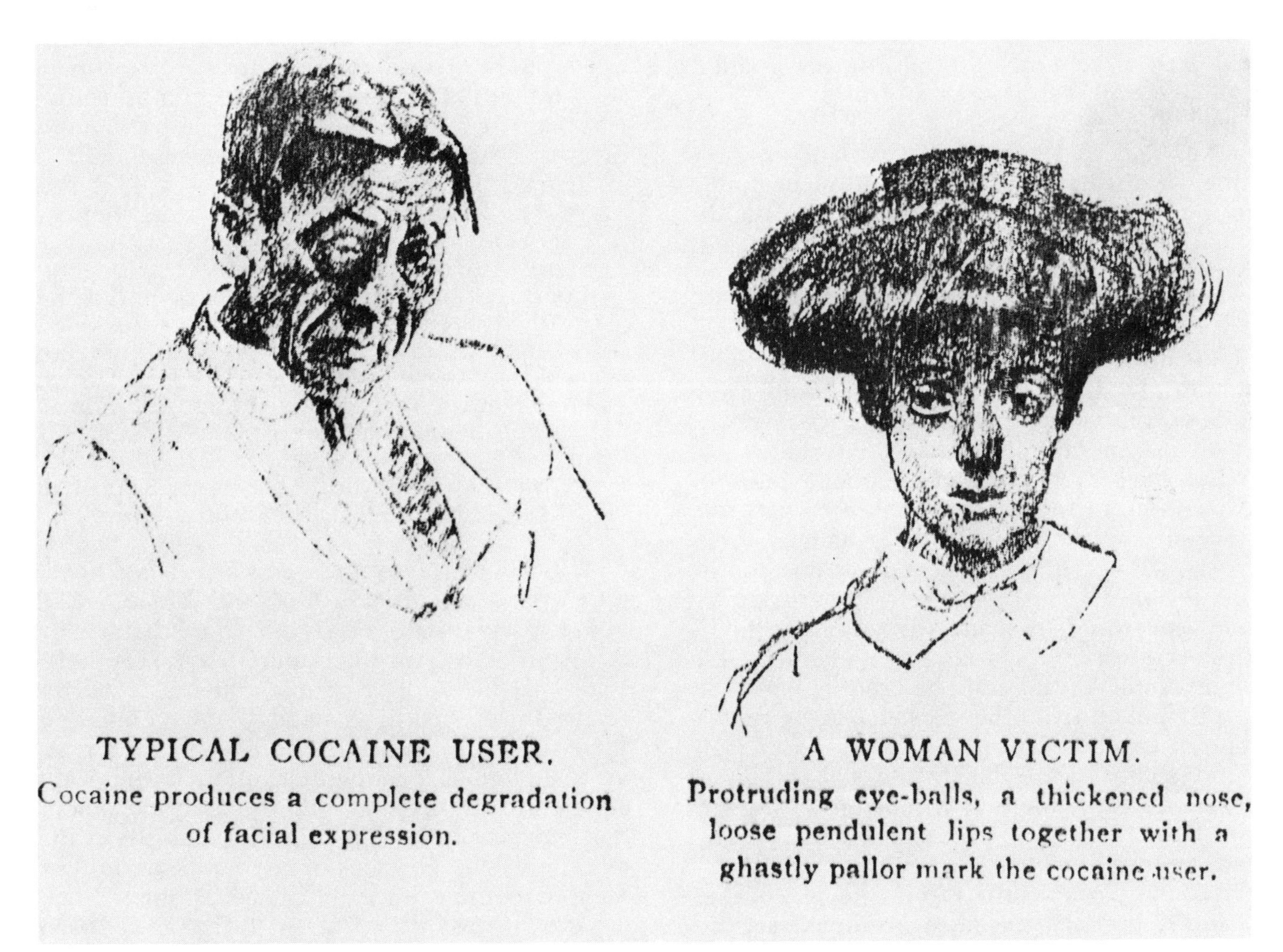


**Fictionalized representations of cocaine users, ca. 1911.** ***(From* Hamptons Magazine *24, no. 1, 1911)***

author, Cleveland Moffet—the man who predicted that automobiles would never become popular in America. Under the bold headline, "Many Crimes Caused By Cocaine," Moffet states: "Let it be noted that crimes are committed by 'coke fiends' not only in their frenzy of desire to get the drug, but in the frenzy of exhilaration that follows taking it. . . . Only a few weeks ago the slayer of little Marie Smith at Asbury Park, New Jersey, confessed himself a victim of the cocaine habit, and no less than the assistant chief of the Chicago Police Department told me of an unsolved murder mystery [the Cleghorn case, January 1910] where suspicion pointed to cocaine." Finally, Moffet solicits some misguided medical testimony in support of his case, quoting Dr. Hamilton Wright, M.D., as follows: "This new vice, the cocaine vice, the most serious to be dealt with, has proved to be a creator of criminals and unusual forms of violence."[10]

The myth that cocaine creates criminals, however, was still not enough to deal a death blow to the drug's reputation. Eventually, all manner of wrongdoing would be associated with cocaine use in an apparent effort to rally the indignation of all types of special-interest groups and prejudicial thinkers.

It was said that cocaine was used to corrupt young girls who were then led unsuspectingly into the white-slave trade. There were, in addition, a host of references to cocaine's abilities to subvert "religious duty" and undermine "the authority of ritual superiors." It was called "the devil's own drug" and was advertised, just as it was in sixteenth-century Spain, as a demonic manifestation from hell. One anthropologically minded critic based his objections to cocaine on a fear of "degenerating back to monkeydom."[11]

When the fear of war and German expansion was beginning to peak, the *New York Times* ran a story on a supposed plot by German agents to import a dangerously addicting cocaine toothpaste into the U.S. According to this article, the toothpaste would be made easily obtainable all over the country until just before a planned German invasion. At that time the product would be abruptly pulled from the market, leaving only a nation of helpless, strung-out cocaine addicts to defend against the attack. In retrospect, this story seems to be more of a conspired effort to generate a fear of cocaine than any actual plot by would-be invaders.[12]

Of all the ploys resorted to by the anticocaine crusaders, perhaps the most insidious was the association they created between blacks and cocaine. Judging from the number and content of "coke-crazed Negro" stories published then, it appears that prohibitionist writers did not hesitate to take full advantage of the ongoing fear and resentment of blacks that had already been established throughout the country.

Whenever blacks and cocaine were mentioned together in the same story, the results of that story were ultimately predictable. The message that cocaine would wrongfully motivate, if not lend superhuman strength to blacks and "cause" them to "rise above their station" appears as a motif in the antidrug literature.

In 1914, Dr. Edward H. Williams cautioned that under the influence of cocaine, "sexual desires are increased and perverted, peaceful Negroes become quarrelsome and timid Negroes develop a degree of 'Dutch Courage'."[13] He then goes on to explain how law-enforcement officers in the South have been forced to increase the caliber of the bullets in their guns to deal with the "cocaine nigger" when he "runs amuck."[14]

If any of his readers had somehow failed to get the message by this time. Dr. Williams endeavored to drive the point home by revealing a sensational statistic—one that was guaranteed to put the fear of blacks and cocaine into the heart of every white American. It is best to quote Dr. Williams at length:

> It seems to be an established fact that the abuse of ardent spirits interferes with good shooting. But such an example as that of the cocaine-crazed negro in Ashville who killed five men dead in their tracks with one shot each, shooting at long range in some instances, demonstrates that cocaine does not impair eyesight or muscular coordination. I doubt if this fact in marksmanship—actually killing, not merely wounding, five men with five shots—has been equalled in modern times.[15]

This same type of message was repeated over and over again. In the *New York Tribune*, Colonel J. W. Watson of Georgia warned that "many of the horrible crimes committed in the southern states by the colored people can be traced directly to the cocaine habit."[16] Dr. Hamilton Wright, who by this time had become an official of the State Department, cautioned the 1910 congressional session with the announcement that "cocaine is the direct incentive to the crime of rape by the Negroes of the South and other sections of the country."[17] Dr. Christopher Koch of the Pennsylvania State Pharmacy Board testified in Washington in 1914 and warned Congress about dangerous cocaine-crazed

# NEGRO COCAINE "FIENDS" ARE A NEW SOUTHERN MENACE

By Edward Huntington Williams, M. D.

## Murder and Insanity Increasing Among Lower Class Blacks Because They Have Taken to "Sniffing" Since Deprived of Whisky by Prohibition.

### Makes Better Marksmen.

### Proof Against Bullets.

### Why Do They Do It?

### From Bad to Worse.

# FOR THE WOULD-BE EXPERT IN AUCTION BRIDGE

By Florence Irwin.

## That Nullos Give Experts One More Chance Against Poor Players Is No Valid Objection to Them.

### Just a Clerk

**Dr. Edward Williams's article in the *New York Times*. (From New York Times *24 June 1914*)**

Southern Negroes. Later on, he said, "Most of the attacks on white women of the South are the direct result of a cocaine-crazed Negro brain."[18]

Dr. David Musto, a modern researcher, made an excellent point when he said, "The fear of the cocainized black coincided with the peak of lynchings, legal segregation, and voting laws all designed to remove political and social power from him."[19] By 1914, Southern blacks found themselves in a depressing situation when they realized that they were to become the initial victims of the anticocaine backlash. Temperance advocates created this unnecessary episode in American history when they began to publicize an unrealistic association between blacks and cocaine in order to capitalize on the nation's prejudices. They believed cocaine's popularity would suffer if people could be convinced that the drug would fuel a black rebellion or, at this time, probably even a single black reaction. For many Americans, this paranoid fear was adequately confirmed by weight of the printed material alone.

At this point, it seemed only right that the nation should mobilize against the evil of cocaine. Because blacks were supposedly so intimately involved with this drug, and because their particular use of it supposedly constituted such a threat to the national well-being, it is not surprising that they were selected as the scapegoat in this situation and made to suffer the ignominious acts outlined by Dr. Musto. Tragically, there is no real evidence to indicate that cocaine use among blacks was any greater, percentagewise, than among whites. There is, in fact, a good chance that the situation was reversed.

This condition notwithstanding, something can be said about the possible origins of black cocaine use in the United States. In 1902, the *British Medical Journal* reported that the drug was actually supplied as a stimulant to Negroes who worked unloading steamboats in New Orleans. Also, some plantation owners were known to have stocked it as they would salt or sugar, and it is said that they rationed it out to black field hands at certain peak times to increase production. The same article mentions that eventually black laborers along the Yazoo River in Mississippi refused to work unless cocaine could be obtained nearby. Ten years later, Dr. Charles Towns wrote: "Overseer[s] in the South will deliberately put cocaine into the rations of [their] Negro laborers in order to get more work out of them to meet a sudden emergency."[20]

This information seems to indicate that the black man's taste for cocaine could have been originally cultivated by the same white people who ended up actually persecuting him on account of it!

The heavy-handed and, at times, fanatical crusade to defame cocaine must be considered successful in terms of the like response it produced in the hearts and minds of the people. Some of the outrageous opinions that surfaced in the wake of this anticocaine propaganda were in direct proportion to the exaggerated notions that prompted these comments in the first place.

An anticocaine campaign worker in Philadelphia echoed the sentiments of many when he said, "If I had my way, I would put them all in asylums where I would keep them until they died."[21] There were also a few people who reasoned that the social harm caused by cocaine counterbalanced all the good derived from it; for these people, the use of this drug could not be tolerated under any circumstances whatsoever! A physician's prescription was considered no adequate justification for the use of this frightful substance and it was even suggested that the loss of the drug to surgery could be accepted if it meant the total elimination of the curse that cocaine had placed on society.

Inexorably mixed with this anticocaine hysteria was the lingering suspicion that the coca leaf was, at best, no less harmful than the straight alkaloid and, at worst, the primary vehicle by which the poison was spread. Coca products were swept up in the cocaine witch-hunt with little thought given to how these two substances differed from each other in potency or potential for abuse. Samuel Hopkins Adams, sneeringly referred to Mariani's wine as a "widely-bruited pick-me-up for lassitudinous ladies,"[22] and legal efforts were being enacted in many states to purge all coca products right along with the ones that contained dangerously large amounts of cocaine.

Some concerned coca supporters rose to defend the leaf against its attackers. Dr. William Golden Mortimer published his massive *History of Coca—The Divine Plant of the Incas,* and in it reviewed nearly every positive aspect of coca consumption. Mortimer's friend and former classmate, H. H. Rusby, also published in favor of coca, always stressing the idea that the whole leaf and the alkaloid were two very different things, each deserving separate treatment. After Rusby finished his coca assignment in South America with Parke Davis and Company, he joined the faculty at Columbia University, where he eventually won the distinction of being named Emeritus Professor of Materia Medica.

The *Paris Medical Journal* ran a number of editorials favorable to coca and Angelo Mariani (possibly in reaction to Samuel Hopkins Adams's comments) offered a thousand-dollar reward for "information leading to the arrest and conviction of any person spreading malicious falsehoods, libelous or defamatory reports intended to discredit the old established reputation of our house or the integrity of Vin Mariani."[23]

Another coca-product manufacturer, C. L. Mitchell, M.D., expressed the frustrations of the coca advocates when he answered a letter from the food commissioner of North Dakota, informing him that his product was no longer legal in that state. He said:

> Dear Sir:
>
> Your favor of September 7 duly received for which please accept my thanks. Owing to the "crank" legislation of many states we have discontinued the manufacture of *all* coca and cocaine preparations. Any "fool" druggist of your state who gets or sells an old package of Coca Bola does it at his own risk, as necessarily, having been put out some time ago there is *no* guarantee, and we will not protect him. The people are getting a little sense into their heads, however *gradually*, and they will sometime realize that prescriptions of both coca and cocaine have an honest and legitimate use by the medical profession. Your state law is silly, and on a par with the nine foot bedsheet laws of Texas and Oklahoma. Of course, your duty is to enforce the law, not to criticize it. I can do that. I am,
>
> Yours very truly,
> Charles L. Mitchell, M.D.[24]

The opinions of Dr. Mitchell and the others went unheard. Beginning in 1913, a rash of state and federal regulations was put into effect that all but completely shut off the supply of coca and cocaine to the American public.

COCA-BOLA Is a masticating or chewing paste made from the leaves of the Peruvian Cocoa plant. A small portion chewed occasionally acts as a powerful tonic to the muscular and nervous system, relieving fatigue and exhaustion, and enabling the user to perform additional mental and physical labor without evil after-effects.

As a remedy and substitute for **TOBACCO, ALCOHOL** and **OPIUM**, in the treatment of those habits, it is invaluable.

**PROF. WM. F. WAUGH, M. D.**, in a paper read before the Pa. State Medical Society (*Phila. Medical Times*, March 1886), calls attention to its value for this purpose. It relieves the insufferable craving for stimulants, and prevents depression, nausea and loss of appetite, and acts as a general tonic, stimulant and sustaining agent. In this last respect it is vastly superior to all tobacco substitutes, etc., which are usually combinations of licorice root and other inert ingredients. It is harmless in its action, creates no habit, and its use can at any time be suspended.

COCOA-BOLA is put up in handsome tin pocket boxes containing sufficient for at least two weeks' use.

**PRICE PER BOX, 50 CTS; BY MAIL 55 CTS.**

**25c** mailed direct to me will bring a **Special Sample Box** with Booklet, "Coca-Bola and its Uses" Interesting and instructive

**C. L. MITCHELL, M. D.**

**1016 Cherry St.,** **PHILADELPHIA.**

**Dr. Mitchell's Coca-Bola. Mitchell wrote a good letter to the Food Commissioner of North Dakota defending coca, but misspelled the name of the leaf twice in his own advertisement.** ***(From Journal of the American Medical Association, 1912)***

By 1914, forty-six states had armed themselves with some sort of anticocaine legislation while only about half that number bothered to do the same for the opiates.[25] Moreover, those states with laws regulating the use of both opium and cocaine often made the penalties for the latter far greater; under the Boylan Act of 1914, for example, the illegal sale of heroin was considered a misdemeanor violation of the New York State public-health law while the illegal sale of cocaine was a felony violation of the New York State penal code.[26]

On March 1, 1915, the United States government began the enforcement of House Resolution 6282, better known as the Harrison Narcotics Act. This law—so far the most comprehensive U.S. effort to control the use of narcotics—was not so much a reaction to the domestic drug situation as it was an attempt to honor international treaty commitments made at The Hague Opium Convention in 1912. Had it not been for the extenuating circumstances brought about at The Hague, it is questionable whether or not Congress would have chosen to test the constitutionality of a law that would conceivably interfere with the rights of personal freedom and the individual's prerogative to pursue happiness. Previously, the Supreme Court felt that any law that purported to restrict the personal consumption of drugs would be clearly inconsistent with the basic concepts outlined in the U.S. Constitution.

The developments at The Hague Opium Convention came to pass in this way. After the Spanish-American War, the U.S. took possession of the Philippines and with it inherited a country with an opium problem far greater than its own. Under the leadership of Bishop Charles Brent, the United States sought to cut off the opium supply to its newest possession by calling for multination cooperation to control the traffic in narcotics.

To this end, the United States convinced the representatives of thirteen nations to gather at The Hague, The Netherlands, in an attempt to work out some comprehensive drug legislation. The American representatives at The Hague were Henry J. Finger, a California pharmacist, the missionary Charles Brent, and none other than Hamilton Wright, M.D., the author of some of the nefarious anticocaine statements mentioned earlier.

Like most international conferences, The Hague Opium Convention was filled with confusion and contradiction. From the very beginning, the English wanted to call off the conference because some essential nations such as Turkey and Peru were not in attendance; the Germans supported this idea by pointing out that thirty-three of the world's forty-six powers were not in attendance and no truly significant world resolutions could be set forth in this atmosphere.

Hamilton Wright managed to contact the British and somehow convinced them to forget their grievance and join the convention. On the day he was to leave for Berlin, the Germans informed Wright that they were now also ready to sit down and talk. As soon as this crisis had been resolved, the Austria-Hungary delegation elected to pull out, reducing the number of attendant nations to twelve.

Undaunted by numbers, or lack of them, the conference got underway amidst a complicated assemblage of varying national interests and special reform issues. Italy, for example, insisted that *Cannabis sativa*, the marijuana plant, be included in the discussions on dangerous drugs; after learning that only the United States would support such a notion, the Italians joined the Austria-Hungary delegation in leaving and that reduced the number of participants to eleven.

Among the remaining nations, there was no shortage of opinion on the subject of what substance should receive the lion's share of the convention's attention. Unlike the Americans and the Chinese, the English wanted to "deemphasize" the opium question and concentrate the convention's efforts on the restriction of morphine and cocaine. This recommendation was no doubt in response to the United Kingdom's richly deserved feelings of national guilt over their role in the Chinese opium problem. This position was given the support of the Portuguese, who had marketable poppy fields under cultivation in Macao and Persia as well as on their own soil. France, The Netherlands, and Russia were all in similar economic situations because of their agricultural interests in *Papaver somniferum*, the opium poppy. The Germans, on the other hand, had no poppy fields in their sphere of influence, but they had many large chemical companies such as Merck of Darmstadt that synthesized much of the world's morphine and cocaine. Needless to say, they were hesitant to support any law that might cut into the profits realized through the export of these two lucrative items. The Japanese, who were just gearing up for the mass production of hypodermic syringes (popular with morphine users) and the manufacture of cocaine from Java-grown coca leaves, joined with the Germans in a desire to center attention on the opium problem or, more accurately, away from the morphine and cocaine issues. Like the English, the Japanese were sensitive about being instrumental in the proliferation of morphine use in China.

In order to reach agreement on at least one issue, the convention recommended that each member nation should endeavor to control its own particular narcotics problem by instituting comprehensive domestic reform measures that would carry international implications. In reply to this suggestion, the Germans raised two very good points. First, they mentioned that most countries already had such laws in effect; next, they asked the United States, which was one of the nations lacking federal control, what guarantee would there be that the U.S. Congress would pass the comprehensive legislation needed to enforce their portion of the convention. The answer, of course, was that there could be no guarantee, but this was not Representative Wright's reply. He insisted that the good faith of the United States government was guarantee enough, knowing all along that it would take a determined if not devious effort to put such a law into effect.

This was a most embarrassing situation for the United States. It quickly became clear that some sort of federal narcotics legislation would have to be forthcoming at home if the Americans hoped to head off the dissolution and failure of the conference they originally instigated and insisted upon having. The conference adjourned in 1912 with the U.S. narcotics law question, and most other considerations, still unsettled.

After the close of the first Hague International Opium Convention (there would eventually be a total of three), Hamilton Wright returned to the United States and began a campaign that culminated in the passage of the nation's first federal narcotics law. Wright selected Representative Francis Burton Harrison to sponsor the bill in the House and, because of his unrelenting support, the law ultimately came to bear the representative's name.

From a comment made by Harrison in 1911, it would be hard to refute the notion that he was a man who had been somewhat influenced by the antinarcotics propaganda of the day. When the question of whether or not coca leaves should be included in a list of dangerous narcotic substances was broached, Harrison argued that they should be because they were used to make "Coca-Cola, Pepsi-Cola and all those things sold to Negroes all over the South."[27]

Despite the indications that sweeping drug legislation may not have been the best thing for the country at that time, President Woodrow Wilson signed Harrison's bill into law on December 7, 1914. Along with the Harrison Narcotics Act came the Narcotic Drugs Import and Export Act, a piece of legislation that was custom-fitted to fill the gap in the American pledges made at The Hague. As far as cocaine was concerned, this law made it illegal to export American cocaine to any country that did not control its own exports.

Ostensibly, the Harrison Narcotics Act amounted to little more than an excise tax that generated the moderate revenue of 1¢ per ounce on all opium, coca leaves, and their derivatives. The law required that anyone who produced, imported, manufactured, compounded, dealt in, dispersed, sold, distributed, or gave away any of these drugs register with the Internal Revenue Service and pay the tax. In addition, they were required to keep accurate records of all the drugs they received or transferred. The law also stated that drugs could change hands only between these registered persons. According to Hamilton Wright, the Harrison Narcotics Act would merely "bring this whole traffic and use of these drugs into the light of day and thereby create a public opinion against them."[28]

After President Wilson made the Harrison Narcotics Act law, however, this seemingly harmless exercise of observing the drug situation through record keeping and tax forms quickly degenerated into a menacing contrivance aimed at the reduction of individual rights and personal freedom.

On the surface, there appeared to be nothing wrong with a law whose main function was to gather information on a medical problem and generate a modest tax in the process. Those who bothered to read the law must have been especially comforted by the fact that there was nothing in it to prevent a U.S. citizen from using any drug nor any provision that would deny registration to any individual who wanted to deal in those substances. As further insurance against alarm, the law also claimed that "nothing contained in this chapter shall apply to the dispensing or distribution of any drug . . . to a patient by a physician, dentist or veterinary surgeon . . . in the course of his professional practice only."[29]

The hidden interpretations of this law would eventually make these constitutional safeguards seem like lies and would also work to accelerate the American drug problem into a full-blown drug crisis.

Agents of the Bureau of Internal Revenue were responsible for much of the trouble when they began to arrest doctors who prescribed drugs to addicts for the maintenance of their habits. The enforcement agents in this case determined that these prescriptions were illegitimate because they were supplied for the mere comfort of the addict population and were not given in an attempt to effect a real cure. In 1919, the Supreme Court ruled that it was illegal to prescribe narcotics for a known addict *(Webb et al.* v. *U.S.),* but at the same time the court hinted that it would be legal to give out drugs to addicts who were registered in government-operated maintenance clinics. Subsequently, licensed clinics were established around the country to care for the addict population. Just three years after the first clinic was opened in Jacksonville, Florida, the Supreme Court ruled (in the *Berrman* case) that it was illegal to prescribe drugs in a maintenance clinic after all. This apparent reversal opened the door for an all-out assault on the clinics by federal enforcement agents.

The arrests on these grounds must have come as quite a shock to physicians who believed they were obeying the letter of the law by laboriously filling out all the regulatory forms that had been handed out by the Treasury Department. One reaction to these surprise enforcement tactics was issued by congressional Representative Lester Volk, who said: "It seems to me that the untutored narcotics agents of this great government . . . might have been better employed than in taking sides in a medical controversy involving the broad subject of what will or will not constitute the proper medication in the treatment of addiction."[30]

An even better criticism of the bureau's activities could be made from the relationship between the disappearance of the last legitimate drug source and the emergence of the corrupt drug pusher. Had the Treasury men allowed addicts and more moderate users to turn to the medical profession to secure their drugs, they could have succeeded in (1) placing every problem drug user under a doctor's care (if only for a brief weekly visit), and (2) realizing the main purpose of the Harrison Narcotics Act by having all drug users registered and placed in an observable subgroup that would be relatively easy to control.

Instead, over aggressive police action succeeded in completely defeating the whole idea of the Harrison Narcotics Act; it forced drug users (and especially addicts) underground, where it was next to impossible to observe or monitor their actions, and it opened the door for unauthorized and unscrupulous drug dealers to enter into the situation and occupy the position that could have been held by trained physicians or, at the very least, government bureaucrats.

Some aspects of the Harrison Narcotics Act reflected the special fear of cocaine that had been so carefully constructed in the previous decade. For instance, the law stated that certain medical preparations containing minimal amounts of opium or its derivatives would be exempted from coverage. No such exemption existed for cocaine preparations, no matter how small the cocaine content was. In the first year after the law was put into effect, there were 1,108 remedies on the market with opium, Can-

**Hamilton Wright, father of the U.S. drug laws.** ***(From*** **New York Times** ***21 June 1918)***

nabis, or chloral hydrate in them, but absolutely none with cocaine.[31]

The numerous amendments that were tacked on to the Harrison Narcotics Act also reflected a high degree of cocaine sensitivity. In 1919, an amendment to the act increased the tax on cocaine and made it illegal to sell, purchase, or dispense the drug except in or from its original stamped package. In 1922, Congress went a step further when it prohibited the importation of cocaine and coca leaves. Actually, cocaine was absolutely banned while there was a partial (and necessary) exemption made for small amounts of the leaf; this was done so there could be at least some cocaine manufactured domestically for surgical purposes. In the same year, federal statutes declared that cocaine was officially a "narcotic" although pharmacologically it has always been, and still is, a stimulant. This is a very important distinction when questions of addiction, abuse, tolerance, and withdrawal are being considered.

If nothing else, the Harrison Narcotics Act and the Narcotic Drugs Import and Export Act combined to give the United States the legislation needed to save face at The Hague. Interestingly, when this temperance-oriented body convened for the third and final time in 1914, Dr. Hamilton Wright was not in attendance. Secretary of State William Jennings Bryant suspected Wright of having a drinking problem after smelling liquor on Wright's breath in several meetings at the White House. Bryant ordered Wright to take a pledge of abstinence that would last until the end of The Hague convention, and when the embarrassed Wright refused, he was immediately dismissed from the U.S. delegation.

This poignant epilogue to the U.S. drug-legislation story is not lacking in poetic justice or moral instruction. On one hand, it seems fitting that Wright's dismissal should have come about as a result of the temperance hysteria that he was so instrumental in developing. However, it can be said that smelling liquor on someone's breath is meager indication of that person's capabilities for leadership or clear thinking. Secretary of State Bryant's actions in this case gave rise to a host of negative and possibly untrue accusations about Hamilton Wright that proved to be completely disastrous to the latter's career.

It goes without saying that this same ill-conceived thinking was applied to the drug situation as well. The attitude that dramatized all the popular fears about drugs (including alcohol) and then carried these alarmist ideas to their logical conclusion was the same attitude that fostered the passage of the Volstead Act (the enforcement arm of the ill-fated Eighteenth Amendment, outlawing alcohol) and the Harrison Narcotics Act.

The enforcement of these acts has always been an unpopular and, to some extent, unsuccessful venture. But more than that, these control methods worked against the best interests of the nation by creating a drug situation that was far more serious than the original problem as it existed at the time the laws were passed.

Some turn-of-the-century temperance advocates lived to see the cocaine problem temporarily fade, then rise again with greater intensity and more complications. A few more of them would go on to see this phenomenon repeat itself at least two more times. Those who have lived through the changes would have to agree that the situation has worsened considerably from the so-called problem days before the anti-drug laws were passed. In any review of the situation, it would be difficult to avoid the similarities between the disastrous effects of the Volstead Act prohibition and the present illicit drug situation vis-à-vis the Harrison Narcotics Act.

In reaction to organized crime's involvement in the traffic and sale of cocaine, the forbidden-fruit factor in the drug's recurring popularity, the dangers of badly adulterated street cocaine, and the amount of misplaced money and effort spent searching out, arresting, prosecuting, and incarcerating cocaine "criminals," it would not be surprising at this point to hear a call for "the good old days" when cocaine and coca leaves were legal and available to use or misuse as people saw fit.

# 12

# Illicit Cocaine in the 1920s: America's Second Civil War

In the 1920s, two contrasting elements in the American character collided over the question of drug use. A pleasure-seeking constituent in the nation went to war against the reactionary establishment to regain the freedom of personal consumption. The subsequent battles waged in the name of this issue created a decade of unrest and national polarization. Fracture wounds suffered in these battles remain in American culture today and have many of the characteristics of those left by the nation's first civil war.

The immediate roots of America's second civil war began shortly after the passage of the Harrison Narcotics Act and the Volstead Act, a time when the country radiated strong feelings of patriotic self-assuredness and ethical superiority that had come as a result of World War I. Americans indulged themselves with the notion that they had won a moral rather than a military victory and they felt that their righteous idealism had triumphed over the evils of European decadence.[1] A segment of the population became flushed with the idea that a puritanical society was a successful one and that Americans had to be pure to be strong. Under this system, hard work and rigorous self-discipline were considered highly desirable traits and were rewarded with social favors. After maximizing the positive effects of these two characteristics, it only followed that this system would scorn and, if possible, punish those who were pursuing idle pleasures or conducting themselves as undisciplined rakes.

This puritanical bent of the twenties was accompanied by a set of equally intense feelings regarding social behavior that were aimed in the exact opposite direction. F. Scott Fitzgerald called this era the "Jazz Age" and his writings reflect the twenties' fascination with glamour, fast-paced social living, and the importance of having a good time. The disciples of the Jazz Age were in open revolt against the morality of puritanism and any prohibitive legislation that might prevent pleasure.

The decade following the end of World War I has also been called the "Roaring '20s" and is remembered as the time of the unconventional flapper and the illegal but romantic speakeasy. These two American institutions represented the nation's alter ego and symbolized its resolve to pursue individual pleasures in spite of any traditionalist reproach or public law that said otherwise.

The pull and haul over the pleasure issue had divided the nation into two great camps. Because of previous developments, however, there were many laws that favored the side of the temperance advocates; at their insistance, these laws became political weapons that were used against Americans who refused to conform to conservative ideals.

The enforcement of these prohibition laws

produced a profound psychological impact on the American people and what they came to believe about drugs and behavior. When narcotics agents began to publish their drug and alcohol arrest figures, the number of offenders in these two categories seemed exceptionally high. Temperance advocates seized on these statistics as proof that their position on the drug issue was a sound one—reminding everyone that they had warned that substances such as alcohol and cocaine would cause crime and breed criminal behavior. It was on the basis of these warnings that they launched their campaign to coerce the government into taking further legal action against drug users. Now, pointing to the large number of "criminals" netted by state and federal authorities after the enactment of the Harrison Narcotics Act and the Volstead Act, they sought to extend the coercion by using the high arrest figures as evidence that there truly was an association between drug users and crime. In this way, temperance advocates hoped to engineer the fulfillment of their own prophecies.

A more appealing logic lies hidden just below the surface of the temperance position. It was not immediately obvious that, since the inception of the drug laws, anyone who had anything to do with drugs or alcohol *was* a criminal, regardless of whether or not he was a harmless, occasional drug user or a moderate social drinker. After 1919, alcohol or cocaine "criminals" were not very hard to find. Surely there were many thousands of otherwise law-abiding Americans caught up in the situation of being made into criminals almost overnight. These people had naturally come to associate crime with acts such as theft, rape, and murder and could not possibly have imagined themselves being classified as real criminals because they drank whiskey or took cocaine. Nevertheless, beginning in the twenties there were wholesale arrests of ordinary citizens who could not muster an ounce of personal guilt over their decision to purchase and enjoy alcohol, and by 1928, a full third of all federal prison inmates in the U.S. were jailed as violators of the Harrison Narcotics Act.

Long before the situation had degenerated to this point, there was some evidence beginning to surface that exposed the temperance position as fraudulent. It suggested that some of their basic assumptions were inaccurate and that prohibitionist zealots may have misrepresented the situation to the Congress by way of the printed word.

Early in the 1920s, the first reliable statistics were released on the number of drug addicts in the United States. A few years earlier, when the drug laws were being written, it was advertised that there were a million and one half Americans addicted to drugs and that eighty-five percent of those people were "confirmed criminals."[2] A religious group known as the Knights of Columbus thought that these figures were much too conservative and warned that the country harbored at least four million dangerous addicts. Then, when the U.S. Public Health Service announced the results of the Kolb-DuMez study of 1923–24, it was revealed that there were no more than 110,000 addicts in the entire nation; the study wisely decided not to guess what percentage of these people might be considered dangerous or members of a confirmed criminal class.

In 1923, nine years after the passage of the Harrison Narcotics Act, Governor Al Smith of New York suspected that there was something wrong with the way the nation's drug problem was handled in the press. He said:

> Agitating the community and increasingly forcing itself upon our attention is the narcotic drug evil. I am convinced that part of the agitation on this subject is due to the sensationalism of some types of newspapers and magazines. Lurid, sensational articles, intended to inflame the imagination of young people and to make the whole subject mysteriously and morbidly attractive, had led to the prevalence of a belief that the use of narcotics is much more general than it really is.[3]

Besides this, there was also growing proof that the new drug laws were ineffective in terms of their ability to stop the spread of cocaine or discourage its use. In fact, the evidence seems to suggest that the passage of the different prohibition laws actually triggered an era of cocaine excess and worldwide distribution.

It has been estimated, for instance, that cocaine abuse in the United States reached an all-time high during prohibition.[4] This information can be supported in theory by observations made on a closely analogous situation in the same country at the same time. It is known, for example, that New York's 15,000 saloons were replaced by 32,000 speakeasies after prohibition.[5] Charles E. Goshen has pointed out that "excessive drinking patterns [in the U.S.] appear to have developed alongside a general hardening of social institutions."[6] These observations beg this question: Is there a causal relationship between the institution of a hard-fisted prohibitive law and a general increase of the use and abuse of the substance that law is trying to prohibit?

During the 1920s, America was not the only country to suffer from this boomerang effect. Other nations laboring over the drug issue at The Hague reported an increase of cocaine traffic and abuse that took place *after* their drug laws were put into effect. In 1921, the Académie Nationale de Médecine in Paris declared that cocaine use in France was increasing rapidly and the laws that were designed to curb it were "ineffective." They said the French anticocaine bill of 1916 was followed by a proliferation of cocaine in central and southern France, especially in Marseille, Nice, and Monte Carlo. The French authorities also noted that the total amount of cocaine seized in the year the drug law was passed was 25 kilograms and that by 1920, that figure had increased to 320 kilograms. They added that in the first five months of 1921, there had been a 100-percent increase over the previous year's total.

The rise of cocaine popularity in Italy is structurally similar to the American and French models. Before World War I, Italy had no significant cocaine problem and no law against it. In 1915, Italy passed an anticocaine law; shortly thereafter, the drug became very popular. Italian authorities have naïvely complained that the growth phenomenon was generated by the arrival of foreign prostitutes who were seeking asylum in their neutral country. Dr. Joseph Gagliano has pointed out that these prostitutes "either exerted enormous influence or had an immense clientele, for cocaine addiction spread rapidly and became regarded as a serious problem."[7]

There was also a great increase of cocaine use in postwar India. Before World War I, the drug was largely confined to urban Delhi, but it soon spread to Benares, Lucknow, Rampur, Ambala, and other places. The movement of the drug around the subcontinent was attributed to the work of trained midwives, shawl merchants, and dried-fruit peddlers. In reaction to the growing use of cocaine, Indian authorities, like their American counterparts, enacted strict control measures that made coca, cocaine, and even ecgonine (a related alkaloid), absolutely forbidden to the general public. Again, like the Americans, the Indians carefully controlled the transfer of cocaine between druggists and the medical profession. As one might expect by this time, the response to these strict prohibitive laws was also remarkably similar in both the U.S. and India. A chemical examiner in the Bengal government said:

> Despite the vigilance of the excise authorities and not withstanding the stringent measures adopted by the government against the sale and possession of this substance by unlicensed persons, there is reason to believe that the cocaine habit has much increased and is spreading rapidly.[8]

Indian cocaine users in the 1920s were dependent on an Oriental connection for their supply of that drug. Japanese-manufactured cocaine (from Java leaves) was smuggled by Chinese sea traders into the port cities of Calcutta and Bombay. The connection must have been a very lucrative one, for Indian customs officials conservatively estimated that, by 1930, there were 500,000 individuals in that country who were taking cocaine habitually for its euphoric effects.[9]

In England, a contributor to *Guy's Hospital Gazette* reported on the effectiveness of that country's anticocaine law of 1916. Dr. Nathan Mutch said that despite official claims that cocaine was on the decrease, his analysis of the situation indicated that "the practice was for many years rife in London in the post-war period."[10]

In 1924, German narcotics agents raided a "cocaine den" in Berlin and arrested close to 100 persons. It was noted that the men and women brought into custody were from all classes of Berlin society and included some high-ranking literary men and university professors. In reference to the country's attempt to enact a workable prohibition law, the New York *Times* reported that German drug authorities and legislators were "in a quandary for a method to suppress the drug at after theatre parties and private homes, where 'snow' is circulated with the same freedom and nonchalance as cigarettes."[11]

Similar reports attesting to cocaine's postwar popularity were submitted from such faraway places as Egypt and Burma and suggest that the world's drug laws did little to contain or arrest the spread and flow of that increasingly popular drug. A German scientist who visited Egypt in 1935 said: "Very strict measures have been taken during recent years against importing and indulging in hashish. Yet there still seems to be a lot of hashish smoking and because of the prohibition, the abuse of opium, cocaine and other narcotic poisons seemed to have increased."[12]

If anything can be said about the prohibition

laws and what effect they had on the drug situation, it would be that they succeeded in giving life to all the fears that had been generated by the propaganda stories of fifteen years past. Now, cocaine and crime were unmistakably associated elements; not because users were transformed into Mr. Hyde–like madmen by the evil magic of the powder, but because being a user became magically synonymous with being a criminal immediately after the laws were signed. To deepen the association, the laws created a situation whereby the drug-using public was forced to deal with criminals to get their cocaine, for its illegality caused it to become the almost exclusive property or smugglers and other unsavory agents.

Turn-of-the-century temperance propaganda also publicized the fact that cocaine adulterated the brain and caused it to deteriorate. Since the Harrison Act, the criminal adulteration of cocaine (to increase the weight and subsequently the price) has probably caused as much pain and trouble among users as the untouched alkaloid. There is, of course, no telling what effect these unnamed adulterants may have had on the behavior of those who unwittingly ingested them along with their cocaine.

Cocaine's illegal status not only affected the drug's tendency to be adulterated but also its susceptibility to ever-increasing price hikes. When cocaine became illegal in France, the street prices for the drug reached astronomical heights and attracted the attention of a genuine criminal class who were eager to realize some easy profits as a result of the newly inflated prices. At this time, a kilogram of cocaine could be purchased in Berlin for 600 francs and easily smuggled across the border and into Paris, where it sold for 15,000 francs. In the United States, a pharmacist could purchase legal cocaine for $2.50 an ounce while illegally it sold on the streets for $142 an ounce and up. In the United Kingdom, before drug legislation, a ten-grain bottle of cocaine in solution sold for 15 shillings. After the laws were passed, the same bottle cost 50 shillings or more.

The transformation of cocaine into a high-priced luxury item had a distinctly noticeable effect on the future course of the drug. As prices continued to increase, the number of people who were able to afford cocaine decreased, until it became known as a drug used primarily by the very rich. As a consequence, the possession of the alkaloid carried with it a certain status that suggested an affiliation with wealth and high society. Through this relationship, cocaine use became regarded by some as the consummate act in the 1920s. The drug's stimulating properties made it the ideal accessory and a must for all drug-minded flappers, highbrows, and other-well-equipped pleasure seekers of the day, whether they could afford it or not.

Those members of the Jazz Age who were deprived of the drug felt the pain of relative deprivation much more than the pain from any supposed withdrawal symptom. They shared those feelings with the millions of Andean Indians who were made to endure a similar situation under the Inca coca-leaf prohibition.

The use of cocaine by the social elite in the 1920s is closely analogous to the preferential use of coca by Topa Inca's touring entourage in the year 1500, and it produced nearly the same response. In both cases an exclusive and admired class of individuals had recourse to a highly desired stimulating drug that was once easily obtained but lately had become hard to come by and strictly controlled. In both instances, the reaction of those who wanted to use the drug but could not was remarkably similar; under the circumstances, they seemed to want it more than ever before and appear to have been ready to take whatever steps necessary in order to get it. The conspicuous consumption and subsequent glamorization of the drug by one segment of the population made the situation worse by inspiring a degree of public envy and encouraging mindless imitation. Finally, and worst of all, it excited a renewal of popularity based on status and gave rise to new and reckless growth.

It can be said that the revival and growth of cocaine popularity in the 1920s began with the enforcement of the drug laws and was fueled along the way by its associations with high living and big money. In a few years, this growth factor became the cause of a great deal of trouble for many cocaine users and stimulated a fresh attack on their life-styles and rights to ingest what they pleased. Although there was a revival of cocaine interest at this time, the 1920s were also the beginning of the end of cocaine's forty-year reign as a drug of preference.

An examination of the fall of cocaine can begin with a look at the glamorous and decadent association between the rich and the drug in the time between the two wars. In 1922, the society-oriented periodical *Vanity Fair* did a story about a gala party at the home of "little Lulu Lenore of the Cuckoo Comedy Co." Her invitations were printed with the campy title "Will You Come to My Snow-Ball?" and her home was said to have been equipped with a miniature drugstore for the occasion. The "special correspondent" who covered the story mentioned that socialite Otho Everard passed out little packages of cocaine and heroin as favors and "kept the company in a roar." The article was entitled: "Happy Days in Hollywood—A social letter from the movie metropolis showing that things are almost what they seem."

In a chic Broadway musical production about high living in the 1920s called *Anything Goes,* lyricist Cole Porter wrote a song entitled "I Get a Kick Out of You." One verse goes:

> I get no kick from cocaine,
> I'm sure that if
> I took one sniff
> It would bore me terrifically too,
> But I get a kick out of you.

Porter himself was known to have hobnobbed with the social cream of his day and is remembered as being one of the more noted eccentrics along the Great White Way. In the 1930s, Porter's tune underwent the type of censorship that comes about when there is a fear that some audiences may be offended by the words of a particular song. When this happens, as it did with "I Get a Kick Out of You," the original lyrics are simply changed to fit the occasion. From time to time listeners heard the first line of Cole Porter's famous song this way:

> Some like their perfume from Spain

or else:

> I get no kick from champagne.

Further connections between cocaine and high society were established in popular fiction. In *À la Recherche du Temps Perdu,* Marcel Proust described the use of the drug by Parisian noblewomen and Aleister Crowley added to the image with his descriptions of Lord Landsend's cocaine machinations over tea in *Diary of a Drug Fiend.*

While the rich were able to cope with the rising price of cocaine and to cultivate their use of the drug, users not so well-to-do found that they could not; if they wished to continue to use cocaine, they necessarily had to go well beyond their means to do so, and therein lies the seeds of several different types of ruination.

This dilemma represents one-half of the trap that cocaine use represented in the 1920s. The victims of this predicament were the weak-willed and the poor, who were engaged in a pitiful, high-priced struggle for possession of a drug they hoped would bring prestige and happiness. It is no wonder, then, that a portrait of the cocaine user as a depraved and desperate character emerges from the postwar literature.

In Pendleton King's one-act play *Cocaine,* Joe, an ex-prizefighter, and his girlfriend, Nora, are perhaps the archetypal victims of the cocaine trap of the 1920s. Their story begins in a depressing one-room apartment in the Bowery. It is 4:00 A.M. and Nora has been out all night picking pockets and prostituting herself in order to raise enough money to purchase a little bit of cocaine. She tells Joe that she has been unsuccessful and that they will have to do without. Joe suggests that it is possible the landlady might give him some cocaine in exchange for sexual favors. At the thought of her lover having congress with another woman, Nora announces that she will commit suicide by inhaling gas fumes. After some discussion, Joe decides that he will die at her side and together they endure not a peaceful suicide by suffocation, but instead, the embarrasment of an abortive attempt. They wake up and discover that the gas that was to fill the room and snuff out their lives has been shut off for nonpayment. Of course, the gas money had already been spent—on cocaine!

Although the story of Joe and Nora is a concentrated and most likely exaggerated example of the horrors of life with (or without) cocaine, there is enough relevant evidence from the real world to make parts of the story uncomfortably believable. It is not known how many men and women became thieves or prostitutes in order to continue an expensive habit they had perhaps acquired in the not too distant past, when drugs were both innocent and cheap. Likewise, it would be impossible to calculate the number of cocaine-related suicides and murders that took place around this time. Nevertheless, it is certain that these kinds of events did take place and, because of the circumstances, probably occurred more frequently in the 1920s than ever before.

Perhaps the most telling account of what cocaine use had become during this period is provided by the Italian writer Dino Segrè. Under the name Pitigrilli, he produced a novel (also called *Cocaine*), about the decadent uses of that drug in Montmartre in the years following the war. The hero of the story, Tito Arnandi, moves in a world of cocaine users and experiences the opposite extremes of underground consumption. When the story begins, he is living in a cheap Montmartre hotel that "fairly reeked with the scent of soap, tobacco, women's perspiration, military leather and the cheap perfumes that invariably saturate all brothels catering to a clientele of small means."

The hotel's only permanent tenants are Tito and a "mysterious man about 50" who dealt a brand of cocaine called "L' Universelle Idole" out of a secret compartment in the stump of his wooden leg. While he was living in this hotel, Tito got a good introduction to the life and ways of the street users and he reports on their condition. In one long descriptive outburst he captures the essence of everything that was wrong with illegal, expensive, and overly glamorized cocaine use:

> They begin, however, with dispensing with all useless expenditures, then they will cut down on the necessary ones, give up their apartments and rent furnished rooms, and from there move into an attic. They sell their fur coats and jewels at ridiculously low prices; then their clothes, and the body next. And they will keep on selling it until someday it will have become so shrunken and wasted that it will be impossible for them to find another buyer. . . . And this is why you can meet certain women, poorly clad and miserable, who just a few months before launched the styles for the well dressed, at Auteuil and Longchamp.
>
> And the fur coat?
>
> Fifty grams of cocaine for it.
>
> And the gold bracelets?
>
> A box that big; but it was all bicarbonate of sodium and penacetin, not cocaine. . . .
>
> Amidst all these human wrecks, half women and half ghosts, struts the peddler with his pockets filled with cardboard boxes of many makes. . . . He does not sell cocaine that is pure; the drug is mixed sparingly with all other ingredients: boric acid, carbonate of magnesia, lactosium. . . . The peddler knows that the addict is well satisfied with any kind of powder that resembles, in a way, cocaine; the important thing for her is to have something to sniff; she does not stop to examine or analyze what she sniffs; during the last stages she won't be able to tell the difference between cocaine and sugar, while during the first stages her interest centers more upon the ceremony attending the sniffing than on the drug itself; what concerns her then is the manner and the way to take it; with the pen point of pure gold perhaps? or an ivory nailfile or a small spade stolen from a saltbox? or maybe the nail of her little finger, trimmed for the purpose?
>
> And the peddler becomes fabulously wealthy in the space of a few months. With an ectar of cocaine he can buy ten thousand francs worth of jewelry and when the clients offer to sell back their empty cocaine boxes, he pays them a cent for every ten.[13]

Later on in the story, Tito comes into a job and a sizable inheritance and begins to move in high circles. He consorts with some rich eccentrics and starts to adopt their ways. Tito is invited to exotic parties where servants grate rocks of cocaine into champagne glasses and South Seas entertainers perform the *danse polynesienne*. The idle rich who gathered at these parties were in a perfect position to fall victim to cocaine's other pitfall. They did not have to worry about economic collapse, adulteration, or seedy peddlers. Their only worries involved the enlargement of their social image and the pursuit of pleasure. Because cocaine was so involved in both of these areas, the people who were able to use this drug on a regular basis often did so, and many of them ruined their lives in the process. They were lured into the other half of the trap represented by expensive and status-enriched cocaine. Unlike the poor, who were frustrated in their efforts to afford cocaine, and those who became poor attempting to pursue it, the rich were ruined by their overuse of the drug that had recently become so exclusive it was thought irresistible.

Tito Arnandi attended a party where cocaine misuse was only one of the gathering's many excessive trappings. It was held at the villa of Madam Kalantan Ter-Gregorianz, located near the Champs Elysées

> in that mundane quarter where the cocaine aristocrats dwell in security. Within the many sumptuous villas where often gather the various *tout* Paris (the *tout* Paris political, the *tout* Paris mundane, the *tout* Paris artistic), one sees many organized parties, who meet and share together the gay ebriety afforded by the drug You can find there the youthful turf and theatrical snobs, the not yet fully pubescent gentlemen who deem themselves duty-bound to exhibit upon their shelves the latest poems launched in the book market, and in their beds the adolescent debutante; and the youthful Parisian boys who have their pajamas designed by artists of the *Vie Parisienne*. They feed themselves with congealed tropical fowl and inject between conversations all the poisons à la mode, the extravagant exaltations, the etheromania, the chlorotomania and the hallucinating white powder from Bolivia. . . . Men and women invite each other to "cocaine parties" just as they would to dinner. In some families the infection extends from the nephew of 15 to the grandfather of 70; cocaine mania for two in many cases; the con-

**Copy of Pitigrilli's *Cocaine*, found by an American GI in the ruins of Adolf Hitler's alpine retreat. *(Courtesy of the Fitz Hugh Ludlow Memorial Library)***

> jugal toxomania is also very frequent, and if the practice would not make the male sexually helpless and the woman sterile, I believe their progeny would be reaching for that white powder the moment they are born. The alcoholized, at least, has the strength to judge the harm he does to himself and can still advise the uninitiated to steer clear of the liquid venom. The cocaine addict, instead, likes to surround himself with proselytes and followers; thus, every victim made by the fatal drug, instead of constituting tangible warning, becomes a veritable hotbed of infection to the novice.[14]

Other writers have commented on the destruction of the rich by excessive cocaine use.

In Proust's last volume of *À La Recherche du Temps Perdu* (*The Past Recaptured* [1928]), the degenerative effects of too much cocaine are evident in the face of the Vicomtesse de St. Fiacre. "Her statuesque features seemed to assure her eternal youth. And besides she was still young. But now, despite her smiles and greeting, I could recognize in her a lady whose features were so chipped away that the lines of her face could no longer be reconstructed. What had happened was that she had been taking cocaine and other drugs for the past three years. Her eyes, circled with deep black rings, wore almost a haunted look."[15]

Aleister Crowley also observed that being able to afford cocaine did not necessarily mean an agreeable relationship with the drug. "We've arranged for a regular supply; but the thing is that the stuff doesn't work anymore. We get insomnia and those things alright but we can't get any fun out of it."[16]

As the cocaine users' plight worsened, so did their reputation. By the end of the decade, it was difficult to find any printed examples of sensible cocaine use. Probably the least dreadful of all the characters that were associated with the drug at this time were jazz musicians and bohemians.

In 1927, Hermann Hesse wrote *Steppenwolf.* In this novel he describes a man, Señor Pablo, who was both jazz musician and bohemian, as well as a user of cocaine.

> No, he said nothing, this Señor Pablo, nor did he even appear to think much, this charming carballero. His business was with the saxophone in the jazz band and to this calling he appeared to devote himself with love and passion. . . . Apart from this, however, he confined himself to being beautiful, to pleasing women, to wearing collar and ties of the latest fashion and a great number of rings on his fingers.[17]

And later on,

> Once when I showed a certain irritation, and even ill humor over one of those fruitless attempts of conversation, he looked into my face with a troubled and sorrowful air and, taking my left hand and stroking it, he offered me a pinch from his little gold snuff box. . . . I took a pinch. The almost immediate effect was that I became clearer in the head and more cheerful. No doubt there was cocaine in that powder. Hermine told me that Pablo had many such drugs and that he procured them through secret channels. He offered them to his friends now and then and was a master in the mixing and prescribing of them. He had drugs for stilling pain, for inducing sleep, for begetting beautiful dreams, lively spirits and the passion of love.[18]

The association that was formed between cocaine and decay in the 1920s was a genuine one. It was also true that many cocaine users were desperate then and that many more pursued offbeat life-styles. Because none of these characteristics is particularly well received by the social mainstream, or by intelligent drug users for that matter, cocaine earned a bad name and began to decline in popularity. The ugly, or at least bizarre image of cocaine use from the impartial literature chronicles this decline, as do the lyrics from the many songs about cocaine composed around this time.

In 1930, the Memphis Jug Band sang "Cocaine Habit Blues":

> Cocaine habit's might bad,
> It's the worst ol' habit that I ever had

and

> Since cocaine went out of style
> You can catch 'em shooting needles all the while.

A traditional ballad from this era called "Cocaine" said:

> Cocaine is for horses,*
> not for men,
> they say it will kill me,
> but they won't say when.

and

> Walked up Fifth, turned down Main,
> looking for a place that sold cocaine.
> There's a sign on the window,
> It ain't no joke.
> The sign on the counter says no more coke.

In 1933, Huddie Ledbetter, also known as Lead-Belly, recorded "Take a Whiff on Me":

> Whiffaree and Whiffarye,
> Gonna keep on whiffing till I die,
> An' ho, ho, baby, take a whiff on me.

Balladeer T. J. Arnall summed up the growing feeling about the drug when he warned

> Lay off that liquor
> And let that cocaine be.

While is not not true that cocaine necessarily has to be the exclusive property of moral decadents and misfits, that's the way it appeared to the only people who left a record of its use and abuse in the 1920s. The musicians and writers at that time saw only the results of cocaine use gone beserk. Their descriptions provide a fairly accurate account of what happened to the people who continued to use cocaine after it became illegal, expensive, and socially powerful.

In the aftermath of their horrifying stories, it was difficult to recall that there were times when the drug was under control and used in a sensible fashion by the respected. It was forgotten that the Andean Indians have enjoyed at least a 5,000-year relationship with their cocaine-enriched coca leaves, and that in the nineteenth century, small coca clubs were formed in Lima and Cerro de Pasco where high-ranking Europeans gathered for harmless recreation. Cocaine critics also lost sight of the fact that between 1863 and 1906, millions of people entered into a benign relationship with Mariani's wine. The testimonials written on behalf of that product demonstrate that cocaine was once associated with scholars, presidents, and popes. Sales records indicate that Vin Mariani found worldwide public acceptance in all segments of society.

Nevertheless, the image of the depraved cocaine user that came about in the 1920s was powerful enough to squash any previous evidence to the contrary. Gone were the lavish endorsements for coca from the cultural elite and the image of the supersleuth who used the drug to sharpen his wits against the forces of evil. Instead, the use of coca or cocaine was used as a convenient explanation for all sorts of "unacceptable behavior." In the late 1920s, for example, a number of German psychiatrists publicized the idea that cocainism negatively affected the libido and gave rise to homosexuality.

*European race horses were often given cocaine to increase their preformance on the track. This problem caused a considerable scandal in Vienna in 1911 and is most likely the reason that cocaine and horses are associated in this way.

# 13

# The Movement to Exterminate Coca: A Case of Botanical Imperialism

The association between cocaine use and disastrous consequences also inspired some Peruvians to link the drug to some of their country's more serious problems. This process began with the publication of *Aveo Sin Nido,* a book about the oppression and miserable living conditions of the highland Indians. The impact of this book on Peruvian social thought has been compared with the American reaction to *Uncle Tom's Cabin.* In a sympathetic but misguided attempt to explain and possibly end the Indians' plight, Dr. Hermilio Valdizan suggested that there was a causal relationship between coca use and racial degeneration. Valdizan's theory was almost completely ignored until the late 1920s, when a strong anticocaine atmosphere was created. At that time, Dr. Carlos Ricketts revived Valdizan's idea, and on November 29, 1929, he proposed a bill to the Chamber of Deputies urging that the coca leaf be eradicated. Like Valdizan, Ricketts mistakenly believed that the ancient Incas had prospered because they prohibited coca use and that their descendants were being brought to ruination because they failed to maintain the sanction. They both reasoned that the elimination of coca would eventually mean the elimination of poverty, disease, and economic instability in modern Peru.

Opposition to the Ricketts plan came from farmers, merchants, and, of course, the Indians themselves. The thrust of their attack against prohibition was that there were no scientific documents stating that coca was any more harmful than coffee or tea, and certainly no sound medical or anthropological literature proving that it caused racial degeneration.

In agreement with the prococa lobby, the Peruvian government voted down the Ricketts bill in 1930. They refused to undertake any sweeping coca reforms until it could be proven that the leaf was a cause of racial degeneration and that its occasional misuse was not merely a symptom of some other factor such as abject poverty or famine.

The attempts to prohibit the cultivation of coca in Peru were given the support of a number of other nations which envisioned a quick end to their own cocaine problems. If it could be proven that coca was harmful, then there was a good chance the Peruvians would dry up the source of the drug, which was so much out of control in Europe and America. If there was no coca, there would be no cocaine and that would be that.

In the 1930s, researchers around the world attempted to give the Peruvians the medical documents they needed to declare the coca leaf a menace; most of them, however, concentrated on the misuse of the cocaine alkaloid and offered only a weak guilt-by-association indictment against the leaf. Besides the German works on cocaine and homosexuality, Italian psychiatrist

Santangelo asserted that cocaine use caused "profligate behaviour, beclouded ethical sense and perversions . . . of sexual desires." The Russian rhynologist Natanson added to the medical literature against cocaine by reporting the presence of nasal lesions and perforations of the septum among heavy cocaine sniffers. By doing this, he refuted an earlier claim that these injuries were caused by neurotic cocaine-charged nose picking rather than the drug itself.

Peruvian medical research in the 1930s was aimed more at coca, but in the end amounted to little more than a further collection of accusations about the plant that lacked a scientific basis. The assistant director of the Institute of Social Hygiene in Lima boldly declared that the majority of the Indians of Peru were addicted to coca and were destroyed by it in body and soul. Dr. Luis Saenz diagnosed the Indian coca chewer as a sick person and claimed that prolonged use produces a true addiction. He said the physical effects of coca use begin with oral deformations and end with a maladjusted endocrine system and basal metabolism. He concluded his study by announcing that coca is an agent of mental retrogression and criminality.

These professional opinions were obviously not enough to bring about the prohibition of coca in Peru, for it is still an ongoing cultural tradition there at this writing. The use of cocaine in Europe and America, however, became very rare in the 1930s, but again, not as a result of the medical testimony amassed against it.

The decline of cocaine popularity was brought about by what could be called a triumph of reason. The pursuit and misuse of cocaine in the previous decade created a sordid atmosphere around the drug that made it unattractive if not repulsive. Because this reputation seemed earned and not merely assigned, cocaine avoidance began to make sense and was practiced by many potential users.

The withdrawal of cocaine from the pharmacopoeia of the confirmed user was eased by the development of amphetamines in 1932. The new synthetic stimulants were enthusiastically received because they provided roughly the same effect as cocaine; in addition, they were legal, cheap, easy to get, and, relative to cocaine, long-lasting. It is certain that many cocaine users disappeared into the ranks of amphetamine users in the 1930s and helped to bring about the apparent decline of the Peruvian powder.

Cocaine popularity was also affected by the severe economic depression that lasted until the beginning of World War II. During that time, the drug, which was already an outrageously expensive luxury item, became even more unaffordable and was driven farther away from those who still insisted upon using it.

After 1935, cocaine references in literature were confined to a few passing remarks about the drug that were usually derogatory. Cocaine use became a term that was used to denote evil associations. The drug was often portrayed in the hands of criminals because, by this time, the mention of the two together villified both.

It was used in this way by detective writers such as Raymond Chandler who wished to establish a disreputable connection for a particular character and define that role as a wicked one. In *The High Window,* for example, Chandler has his villains associated with "coke peddlers" and other obvious wrongdoers. Dashiell Hammett also relied on the drug's bad reputation, and probably added to it, when he suggested cocaine was the perfect drug to rob a bank on in *The Red Harvest.*

Possibly the most damaging piece of anti-cocaine information to come out in the 1930s was the drug's reported association with the Nazi high command. Luftwaffe Commander Hermann Goering was said to have lost eighty-five pounds while dieting on cocaine and the Fitz Hugh Ludlow Memorial Library in San Francisco has a copy of Pitigrilli's *Cocaine* that was found in the ruins of Adolf Hitler's retreat in Berchtesgarten. It has been suggested that excessive cocaine use was a key factor in Hitler's well-known paranoid fits and his delusions of power and greatness.[1]

By the outbreak of World War II, cocaine had become such a rare and esoteric vice, it was no longer considered a topic worthy of much attention. The only current cocaine-related problem left to battle was the use of the leaf by the Andean Indians.

Under pressure from different sources, the Peruvian government instructed the Department of Pharmacology of the National Institute of Hygiene in Lima to investigate the coca situation and recommend an official course of action. This directive stimulated a great debate on the psychological effects of coca and its role in the lives of the Indians. A psychiatrist named Gutiérrez-Noriega took a stand on this issue very close to the one set forth in the Ricketts' Bill. He believed that coca chewing was destroying the Indians because it was toxic and because it created a negative effect on their intelligence that doomed them to mediocrity. In 1948, he and Zapata Ortiz administered the Binet-Simon

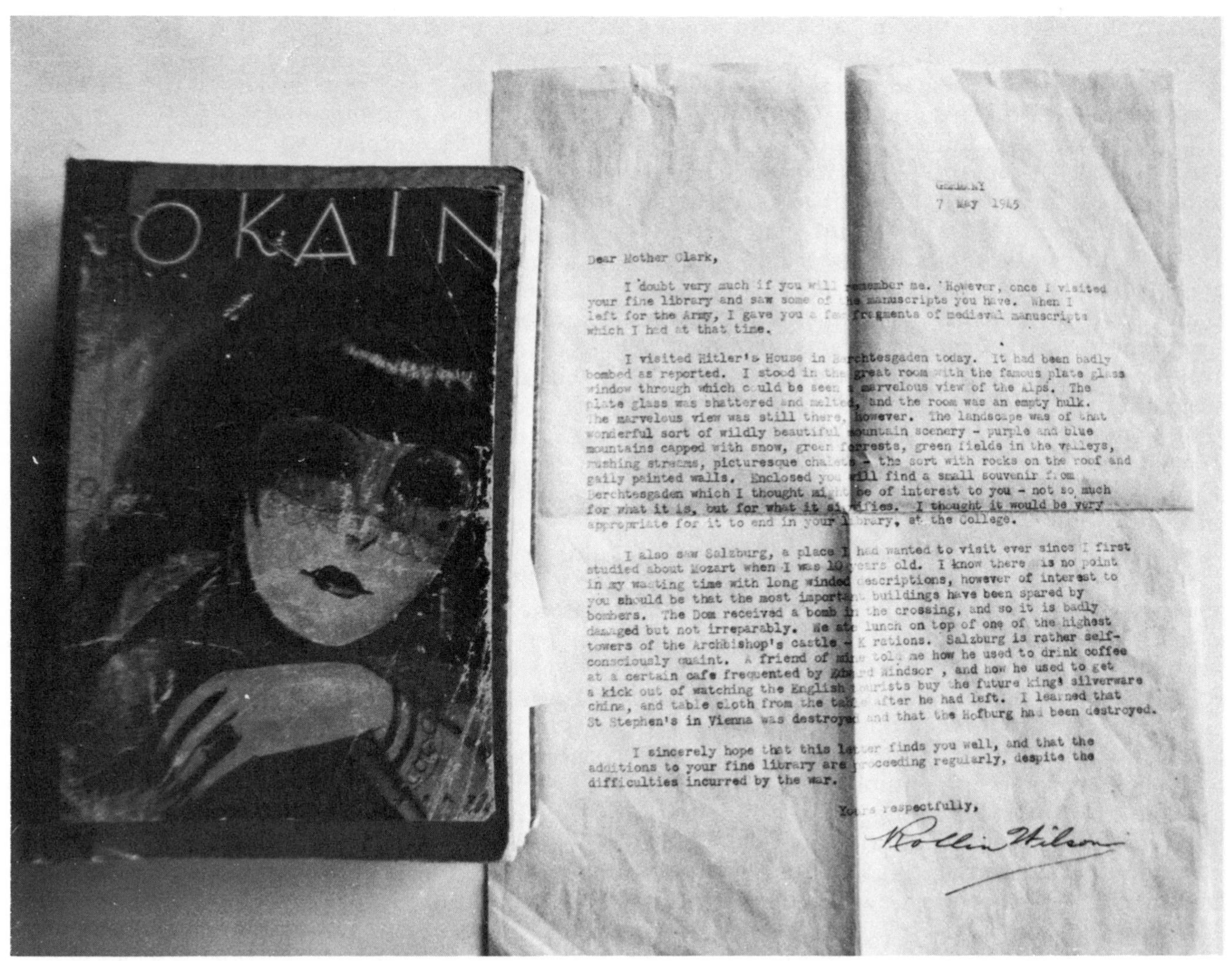


GERMANY
7 May 1945

Dear Mother Clark,

I doubt very much if you will remember me. However, once I visited your fine library and saw some of the manuscripts you have. When I left for the Army, I gave you a few fragments of medieval manuscripts which I had at that time.

I visited Hitler's House in Berchtesgaden today. It had been badly bombed as reported. I stood in the great room with the famous plate glass window through which could be seen a marvelous view of the Alps. The plate glass was shattered and melted, and the room was an empty hulk. The marvelous view was still there, however. The landscape was of that wonderful sort of wildly beautiful mountain scenery - purple and blue mountains capped with snow, green forrests, green fields in the valleys, rushing streams, picturesque chalets - the sort with rocks on the roof and gaily painted walls. Enclosed you will find a small souvenir from Berchtesgaden which I thought might be of interest to you - not so much for what it is, but for what it signifies. I thought it would be very appropriate for it to end in your library, at the College.

I also saw Salzburg, a place I had wanted to visit ever since I first studied about Mozart when I was 10 years old. I know there is no point in my wasting time with long winded descriptions, however of interest to you should be that the most important buildings have been spared by bombers. The Dom received a bomb in the crossing, and so it is badly damaged but not irreparably. We ate lunch on top of one of the highest towers of the Archbishop's castle - K rations. Salzburg is rather self-consciously quaint. A friend of mine told me how he used to drink coffee at a certain cafe frequented by Edward Windsor , and how he used to get a kick out of watching the English tourists buy the future king's silverware china, and table cloth from the table after he had left. I learned that St Stephen's in Vienna was destroyed and that the Hofburg had been destroyed.

I sincerely hope that this letter finds you well, and that the additions to your fine library are proceeding regularly, despite the difficulties incurred by the war.

Yours respectfully,

Rollin Wilson

**Copy of Pitigrilli's *Cocaine*, with letter.** ***(Courtesy of the Fitz Hugh Ludlow Memorial Library)***

Intelligence Test to 350 coca chewers and found that their mean IQ was 67 and that their average mental age was six years, seven months. In addition to this, Gutiérrez-Noriega and his students maintained that coca chewing helped to promote malnutrition because undernourished Indians chewed coca leaves to relieve their hunger pangs instead of eating food.

Dr. Carlos Monge, the director of the Institute of Andean Biology, refuted Gutiérrez-Noriega's findings and came to represent the other side of the coca debate. To begin with, he argued, coca is not toxic to the Indians because their average daily consumption is not great enough to cause them any harm. In defense of this, he pointed out that even though the Indians chewed approximately 60 grams of leaves a day, their bodies only process between 0.05 and 0.09 grams of cocaine. Not only did Monge refute the notion that coca was harmful to the Indians, he went beyond what was expected and declared that coca actually benefited them because of their poor diet and adverse living conditions. He proposed what has been called the "Andean Man Thesis," which posits that coca use is justified for those living in high altitudes. He came to this conclusion by noticing that coca chewing in Peru is almost universal between 12,000 and 15,000 feet above sea level, that the number of chewers declines rapidly between 12,000 and 8,000 feet, and that below 8,000 feet the practice is virtually nonexistent.

Peruvian pharmacologist Fernando Cabieses Molina also criticized the work of Gutiérrez-Noriega and his students. Molina rejected their IQ experiments because they were conducted on prisoners, mental patients, and juvenile delinquents who often used pure cocaine in addition to coca. He wondered why normal, socially acceptable coca-chewing Indians were overlooked

**Adolf Hitler's bookplate as it appeared in his copy of Pitigrilli's *Cocaine. (Courtesy of the Fitz Hugh Ludlow Memorial Library)***

when they represented the vast majority of users. His main objection to Gutiérrez-Noriega, however, centered around the fact that cocaine is slowly and harmlessly absorbed by the digestive system when it is orally ingested in small amounts as it is in the coca-chewing process. In his article entitled "The Anti-Fatigue Action of Cocaine and Habituation to Coca in Peru," he said:

> Many attacks on the coca habit in Peru have failed to take into account the concentration factor. Concentration is not dosage; the concentration of a drug at the tissue level, where it becomes effective, depends not only on the amount of the drug ingested, but also on the route of administration and the body's ability to absorb, destroy, and eliminate it.
>
> It has been shown that the minimal daily dose of cocaine as it appears in the coca leaf chewed by the *coquero* . . . [is] considerably below the toxicity threshold of cocaine, even if it is injected directly into the blood stream.[2]

The movement to restrict coca chewing in Peru suffered a serious setback in 1950 when Gutiérrez-Noriega was killed in an automobile accident. In December of that year, the Commission of Inquiries declared that Peru would be willing to suppress the coca habit but recent investigations had suggested that the leaf might be a factor in acclimatization at high altitudes. The government was understandably hesitant to restrict the use of the leaf until some conclusive results of this association could be established.

All of this suddenly changed in March 1951 when the United Nations sent a mission to Peru to "gather information" on the suppression of the leaf. After hearing the official government position on prohibition, the mission impatiently advised that it was not really necessary to have all the scientific facts completely answered before beginning a trial prohibition in some communities.[3] For the next three years, the Peruvian government undertook a series of minor steps and pilot programs that were aimed at pleasing the United Nations while not completely giving in to their recommendations for absolute eradication of the leaf. Then, in 1954, the Peruvian representative to the Economic and Social Council unexpectedly announced that his country had decided to agree with the UN and join with it in an active denunciation of the coca leaf as a detri-

**Both Hitler and Goering are suspected of being cocaine users. *(Courtesy of Wing Commander Asher Lee)***

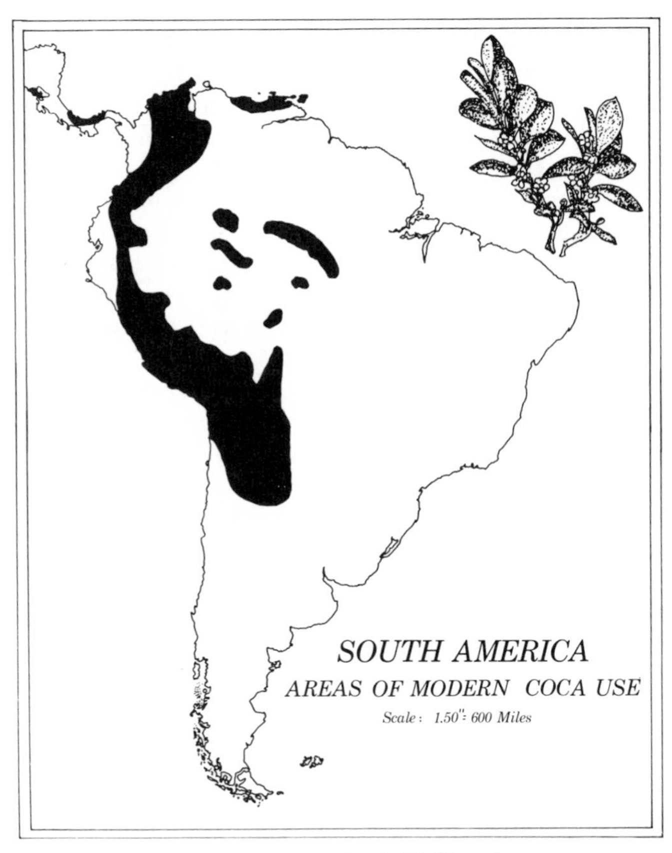


**(Courtesy of John A. P. Kruse)**

mental substance. This was the first time in modern history that Peru had bowed to outside pressure.

In 1956, no new licenses were given to grow coca and the Institute of Inter-American Affairs began a program of so-called agricultural improvement that was designed to replace coca with other crops such as peanuts. Soon after this, Cornell University sent technical advisors to Peru to "assist the Indians in their adaptation to modern living."[4] Included in this project was an effort to change many of the Indian traditions, including coca chewing. In 1958, the Peruvian government raised the duties of coca imports from Bolivia and supported the Third Draft of a Single Convention on Narcotic Drugs, which proposed to permit coca chewing "temporarily for a maximum of twenty-five years."[5]

The abrupt shift in the official Peruvian attitude toward coca chewing can be explained in terms of its adaptability to the political wishes of non-Andean peoples. Since the 1925 International Opium Convention in Geneva, the United States government had been trying to solve its own cocaine problems by insisting that other countries, such as Peru, limit or eliminate the cultivation of coca. Shortly after World War II, the United States made another push in this direction; it was prompted to do so by reports that Peru had significantly increased its production of coca leaves and crude cocaine to make up for the destruction of the Dutch East Indies coca fields by the Japanese. U.S. drug officials were worried about what was termed "an overabundance" of coca and cocaine in Peru and they nervously watched for any signs that might indicate a renewal of popularity in North America.

In April 1949, federal narcotics officers found what they were looking for when they discovered two pounds of Peruvian cocaine in the diplomatic pouch of Cuban Consul Rafael Menacho Vicente. U.S. Narcotics Commissioner Harry J. Anslinger detailed the particulars of the Vicente arrest to all the new services, and the press labeled the event a scandal. The story received much publicity in the United States, partly because it broke on the heels of an embarrassing diplomatic encounter with the Cubans. Only two weeks before the Vicente arrest in New York, three drunken U.S. sailors "befouled" the statue of Cuban hero José Martí and stirred some unpleasant anti-American sentiments in Havana. The news of a Cuban diplomat's being caught smuggling cocaine into the United States was relished as a political counterpunch deluxe; in addition, extensive coverage of the story provided Commissioner Anslinger with an opportunity to broadcast a stern warning about cocaine that was reminiscent of some turn-of-the-

**Coca in Altiplano market. (Courtesy of Bavaria-Verlag)**

century temperance propaganda. A *Time* magazine article about the arrest said: "In Washington, Commissioner Harry J. Anslinger reported that the U.S. was swamped with the biggest influx of cocaine in 20 years."[6]

Anslinger knew best of all that this was not saying much, for since the early 1930s, cocaine had not been a very popular drug and the import totals for the United States could not have been very high. Nevertheless, Anslinger had a point to make about the drug Vicente carried, and it came across clearly later on in the *Time* article: "Where it came from was no big problem; the source, said Anslinger, was unquestionably Peru."[7]

Anslinger's reference to Peru carried the implication that this coca-chewing and -producing nation was somehow responsible for the irresponsible abuse of the cocaine alkaloid in the United States. It is probably no coincidence that the United Nations' efforts to prohibit coca production began (with considerable assistance from the U.S.) just eleven months after Anslinger's comments.

The idea of ending the cocaine problem by exterminating the coca plant has always had a tremendous appeal for law-enforcement officers. Even though cocaine had not been a problem anywhere for the previous twenty years, there was a great temptation to make sure that it could never resurface as one again. When the United Nations turned its attention to Peru and coca chewing, it is difficult to believe that it did so out of a purely altruistic concern for the health and well-being of the Andean Indians. Its concern over this Peruvian internal affair is highly questionable in light of its concern over the extra-Peruvian abuse of cocaine. In this matter the United Nations spoke for every nation that was willing to sacrifice the rights of the Indian as well as Peruvian sovereignty in pursuit of its own motives. In doing so, it compounded the severity of the crime by using the health of the Indians as an excuse to eliminate the best evidence available that cocaine is harmless in moderation.

**Harry J. Anslinger, longtime drug commissioner in the United States.** ***(From* Bulletin on Narcotics *4, no. 2, 1953)***

Harry J. Anslinger's warning about the United States' being swamped with a large influx of cocaine was about as accurate as his discreditable medical opinion concerning marijuana use. In 1960, more than a decade after he made that statement, (that the U.S. was swamped with the largest influx of cocaine in twenty years) the total amount of cocaine seized in the U.S. by all law-enforcement agencies was only six pounds.[8]

It is generally agreed that cocaine played an insignificant part in American culture in the 1950s and was rarely seen. Even groups that had known associations with the drug seemed to have been estranged from it during this time. In 1954–55, for instance, 409 jazz musicians in New York City were interviewed about their use of drugs. Although 54 percent of the musicians admitted to occasional marijuana use, and 24 percent of them to the occasional use of heroin, "only a few" said they were using cocaine.[9] The 1959 edition of the *Pharmacological Basis of Therapeutics* lent support to these findings by declaring that cocaine was currently "uncommon" in Western countries. Considering these facts, it is difficult to understand why the U.S. Congress chose this time to increase the penalties for cocaine use. Judging from its reaction, one would get the impression that the nation was in the middle of a serious cocaine epidemic. In 1951, the penalty for failing to register as a cocaine distributor became identical with the penalty for smuggling large sums of cocaine into the country; in the same year, judges were required to sentence cocaine users to mandatory prison terms.[10] In 1956, the mandatory prison terms were increased with the passage of the Narcotics Control Act, a law that could also sentence a cocaine trafficker to life imprisonment or death.[11] There was also a stipulation that denied eligibility for a suspended sentence, probation, or parole even on the first conviction.[12]

If there was any significant movement of illicit cocaine operating under the shadow of the Narcotics Control Act, it was a well-kept secret. Throughout the 1950s and for the better part of the 1960s as well, there were virtually no large shipments of cocaine intercepted at any U.S. border and only a minimum of anticocaine rhetoric from law-enforcement officials. As late as 1967, President Lyndon Johnson's commissioner of law enforcement and administrator of justice said that cocaine traffic "was not a matter of concern."[13] Because the total amount of cocaine seized in the entire country that year was under fifty pounds, the commissioner's statement seems perfectly justified.

Indeed, during the 1950s and the first half of the 1960s, U.S. drug officials had little to worry about and a lot of time on their hands. Responding to the possible threat of wholesale drug and alcohol abuse before 1920, the United States government gave birth to many new branches of law enforcement and appropriated enough funds to support a small army of agents. With the relegalization of alcohol behind them and the nation's drug problem in reasonable check, these agents and agencies were able to occupy themselves in some international projects, such as trying to limit the production of opium in Turkey and the coca plant in Peru and Bolivia.

The efforts to kill off the cocaine giant while it was asleep and away in South America were renewed during this time. In 1962, the U.S. sent a representative to the United Nations Commission on Narcotices and (in the great tradition of selecting men such as Hamilton Wright for important jobs of this nature), Harry J. Anslinger got the call. It is not surprising, then, to learn that in the same year, the United Nations Permanent Central Opium Board and the UN-related World Health Organization began another campaign to eliminate the coca plant in South America. There can be little doubt that Anslinger was instrumental in giving new life to the old U.S.-inspired idea of practicing botanical imperialism with regards to the cultivation of drug plants in other nations.

# 14

# The 1960s and 1970s: "Cocaine Don't Drive Me Crazy"

It is singularly appropriate that while Harry J. Anslinger was at the United Nations pushing for the demise of the "dangerous" coca plant (and, by extension, cocaine), he issued a statement to the press saying that "cocaine abuse is not at present a problem in the U.S.A."[1] Had he made that statement after the Vicente arrest in 1949, he would be remembered as a man who had his hand on the pulse of the nation's cocaine situation—he would have implicitly made reference to almost a decade of cocaine inactivity in the 1940s and have correctly anticipated another that would last through the 1950s.

Representative Anslingers's remark at the UN in 1962 is appropriate only because it is so much in keeping with other notoriously incorrect statements he has made about drugs. In 1949, Anslinger declared that the country is being "swamped" with cocaine when, in fact, the U.S. was in the middle of its greatest period of cocaine abstinence ever! In 1962, Anslinger declared that there was no cocaine problem in the United States when, in fact, the drug was just on the verge of an unbelievably popular explosion.

The reasons for cocaine's renaissance in the second half of the 1960s are not entirely clear. Different authors have suggested everything from a general rise in promiscuity to the near simultaneous rise in marjuana use as explanations for the cocaine phenomenon. It has also been suggested that government crackdowns on illegal amphetamine operations around this time did for cocaine popularity in the 1960s what cocaine disenchantment did for amphetamine popularity in the 1930s. Because there is never an adequate single explanation for any complex event, it is likely that all the factors mentioned above, and many more, contributed to the resurgence in the popularity of cocaine in the 1960s.

Although there was surely a slow but steady increase of cocaine use beginning around 1965, the drug's bad reputation muted much of the evidence that would normally have accompanied this process. The individuals who rediscovered cocaine in the mid-1960s were understandably hesitant to publicize their find because they feared being labeled as users of "hard drugs." In addition, the Narcotics Control Act's threat of life imprisonment or death for cocaine users encouraged early experimenters to maintain a very low profile and keep the news of cocaine's growing popularity to themselves.

These factors combined to set the stage for the drug's first important public appearance in the 1960s. The event was the 1969 release of a film called *Easy Rider*. In the beginning of this story, two young Americans enter the La Contenta Bar, a sleazy roadhouse on the Mexican border. Before long, they are out back sniffing some white powder off a broken mirror and negotiating for the purchase of a large cache of the stuff.

While the substance they wanted to buy is

never explicity identified, an exchange between Jésus, the Mexican drug entrepreneur, and Wyatt, one of the American purchasers, leaves little doubt that the white powder on the mirror is cocaine:

JÉSUS: Está mal, Hermano. Por la vida, Hermano. [It's not good, Brother. For the life, Brother.]
WYATT: Sí, pura vida. [Yes, it's pure life.]

Another motion picture featuring cocaine use, *Night of the Following Day,* was released a full year ahead of *Easy Rider* but enjoyed none of its success. In *Night of the Following Day* old drug stereotypes were reinforced through some familiar associations. This time felons flamboyantly sniffed cocaine through rolled-up 100-dollar bills while in the act of committing a real crime (in this case, kidnapping). As one critic observed, "The film can be said to have contributed little if anything to 'cocaine consciousness.'"[2]

*Easy Rider,* on the other hand, broke through this old drug stereotype by making heroes out of Wyatt and his companion. After they purchased a large amount of cocaine at the La Contenta, they smuggled it back to the United States and immediately sold it. The money from the deal is used to finance a picaresque adventure across the country that ends when the two likable cocaine merchants are martyred by "dangerously uptight, violent law-and-order Americans."[3]

The alternative view of drugs and drug users provided by *Easy Rider* was eagerly devoured by the film-going public, and the production turned out to be an enormous box-office success. On a larger scale, the success of *Easy Rider* reflected a change in the attitude of an entire generation. Traditional values were beginning to be challenged and people felt that they could no longer be content with outdated and ineffective solutions to urgent problems. The 1960s were a time of revolt, dissent, and reevaluation. In terms of the drug situation, and specifically in reference to cocaine, these attitudes and emotions would come to play a very important role.

By the time *Easy Rider* was released, there was already a sizable number of Americans who had discovered that they could smoke marijuana or sniff cocaine and not suffer any of the disastrous consequences they had been told about.. At this point, the carefully constructed fifty-year crusade against certain drugs began to take on the appearance of a gigantic hoax. As the realization of an organized deception started to take hold, a degree of hostility toward the drug laws, the government that created them, and the agents who enforced them began to set in. It was a clear-cut case of the "chickens coming home to roost."

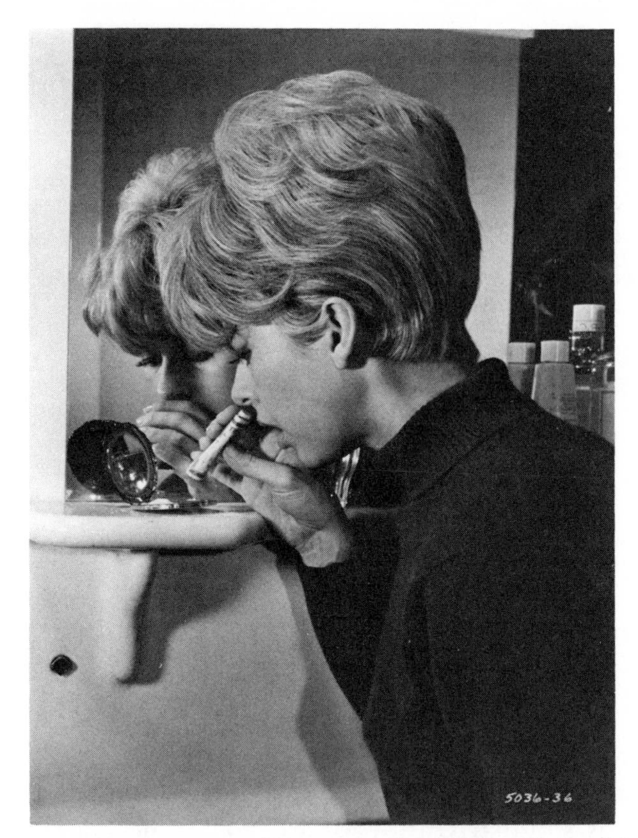

**Actress Rita Moreno sniffs cocaine in the film, *Night of the Following Day. (Courtesy of Universal Studios)***

Old Federal Bureau of Narcotics–sanctioned propaganda films such as *Reefer Madness* and its cocaine counterpart, *The Pace That Kills,* were revived at this time and shown on college campuses; this time around, however, they were perceived as being absolutely ridiculous and only served as a tragicomic reminder of the government's exaggerated, emotional, and essentially wrong position on drug use.

The time was right for the drug-control issue to join the Vietnam War, racial discrimination, and political corruption as a subject on which it was easy to take sides. At a time when people were sensitive to issues such as civil rights and discrimination on the federal level, the marijuana and cocaine stories must have found many sympathetic ears. In the late 1960s, protest marches, draft dodging, refusal to vote, and drug use all became accepted forms of protest against the system, and any one of these activities was enough to secure membership in a growing counterculture that sprang up around dissatisfaction with government policy. While drug use found an additional and popular avenue back into society through its association with this counterculture movement, there were also many disastrous attitudes formed about drug use by this avenue that were as far to the left of center as Harry Anslinger's were to the right.

By 1970, there was a number of signs indicating that cocaine use had once again become part of American culture. If the amount of cocaine seized in police raids and custom inspections can be used as a rough estimate of the amount of illicit cocaine traffic in the nation, and the rate at which this figure grows, then it is interesting to know that there was a more than 150-percent increase in cocaine confiscations between the years 1967 and 1970.*

In response to these figures, the United States government decided to repeal all existing federal drug legislation and replace the former laws with the Comprehensive Drug Abuse Prevention and Control Act of 1970. This act listed cocaine as a "Schedule II drug," or one with a currently accepted medical use but a high potential for abuse that can lead to severe physical or psychological dependence.[4] Also under this act, cocaine was once again mistakenly termed "a narcotic."

The Comprehensive Drug Abuse Prevention and Control Act of 1970 created some new offenses that suggest the U.S. lawmakers were still mindful of the supposed South American influence on North American drug abuse. Section 844 declared that it was now a federal crime for anyone to manufacture or distribute a Schedule II drug anywhere in the world knowing that it would be illegally imported into the United States. This means that a Peruvian or Bolivian cocaine manufacturer can be charged with a U.S. federal crime if he is suspected of diverting his product into illegal channels, and if jurisdiction over him can be obtained.

Another closely associated indication of reemerging cocaine popularity came from the United Nations. In 1970, it decided to create a special fund to help reduce the agricultural production of raw drugs. Operating under the delusion that this effort would somehow help destroy drug-abuse motives in North America, the United States promptly poured two million dollars into the project. After nearly a decade, this fund can claim responsibility for only a partial one-crop suspension of opium-poppy production in Turkey and virtually no limiting effect on the production of coca in South America. There are, in fact, estimates that the South American coca crop is increasing by about ten percent a year.* Despite the combined efforts of the United States and the UN, cocaine consumption in the U.S. continued to grow, as the *New York Times* said, "by leaps and bounds." By 1973, the amount of cocaine seized in the United States totaled just over 1,300 hundred pounds, and by police admission, that figure represented only about one-tenth of the actual total.

Perhaps the greatest evidence of cocaine's resurgence, however, is its reappearance in the music, literature, and films of the early 1970s. During that time, more than a dozen popular songs mentioning cocaine were released by different recording artists. One of these songs, a tune entitled "Champagne Don't Hurt Me, Baby" by folk singer Eric Von Schmidt, was actually released in the early 1960s, but had to wait nearly a decade before it became popular. By 1972, the attitude of the underground drug culture had caught up with Von Schmidt's prophetic lyrics and embraced them as representative of their new convictions. The rediscovered lyrics said:

> Champagne don't hurt me, baby,
> Cocaine don't drive me crazy,
> Ain't nobody's business what I do.

And later in the same tune:

> Drink your whiskey, drink your wine,
> Narco boys gonna treat you fine,
> And it ain't nobody's business but my own.

Moviegoers in the 1970s were made aware of the cocaine boom by the drug's appearance in a number of celebrated film productions. After thirty years of invisibility (as far as motion pictures were concerned), cocaine use suddenly became a component in many screenplays and was used as a device to communicate all sorts of messages. In the film *Willie Dynamite,* as in *Night of the Following Day,* cocaine was used by wrongdoers for evil purposes; in *Gordon's War,* cocaine abuse triggered a violent vigilante action. The film

*The US Department of Health and Human Services reported that in 1980, one third of all Americans between the ages of eighteen and twenty-five had tried cocaine. In 1962, according to the same agency, the figure for that same age group was barely three percent!

*According to the National Coca Enterprise of Peru, in 1977 over 17,000 growers were authorized to produce 10,450 tons of coca leaves on 64 square miles of land. Reliable but unofficial sources contend that all these figures should be roughly doubled.

*Papillon* had escaping prisoners chew coca leaves for endurance. The hero of *Superfly* made his fortune dealing in cocaine; in *Annie Hall,* cocaine ignorance was used to get a laugh. Sherlock Holmes's use of cocaine was capitalized on nearly 100 years after the fact by such films as *The Private Life of Sherlock Holmes* and *The Seven Percent Solution.* The drug also made cameo appearances in the films *Turkish Delight, The Discreet Charm of the Bourgeoisie, Looking for Mr. Goodbar, A Star Is Born, Pretty Baby, Blue Collar,* and many others.

Evidence of cocaine's newfound popularity in the 1970s is equally represented in the publishing world. A golden coke spoon appeared on a cover of *Esquire* magazine in 1971 and two years later, *New York Magazine* did a feature on the drug entitled, "The Big New Easy Entry Business"; in that article, cocaine was referred to as "*the* drug." Shortly after this, the Washington *Star-News* ran big cocaine stories three days in a row under the lead "Cocaine: The New Number One Drug." *Rolling Stone* did two cocaine features in as many years and a cocaine story even appeared in *Newsweek* under the "Life and Leisure" section. By the end of the 1970s, virtually every major newspaper and news magazine in the country had featured cocaine in a number of issues.

In response to the need for information about this hitherto forgotten drug, several books appeared on the market attempting to provide answers for often-asked cocaine questions. *The Gourmet Cookbook* by White Mountain Press and *Cocaine Consumer's Handbook* by David Lee were compiled primarily as reference sheets for street users and dealers. These books cover topics such as how to run simple tests for purity, the legal consequences of cocaine use, and a guide to cocaine terminology and paraphernalia. The success of these two publications can only be attributed to those Americans whose interest in cocaine is practical rather than theoretical. In the same vein, a monthly magazine called *High Times* carries current cocaine stories of every description. In this slick publication, illicit street drug prices are quoted and readers are informed of the latest arrests following the regular column called "Cocaine Confidential." There is a decided procoaine flavor to all their stories and throughout the magazine, drug use in general is treated in a conspicuously casual way. Keeping in mind all the previous antidrug literature, *High Times,* the so-called Magazine of High Society, seems to be proof that for every action there is an equal and opposite reaction.

On the other hand, cocaine popularity in the 1970s occasioned the arrival of some serious books on the subject that have answered some of the more important cocaine questions. The first of these was the 1971 publication of Richard Woodley's *Dealer: Portrait of a Cocaine Merchant.* In *Dealer,* Woodley presents a sober, insightful, and in all probability very accurate account of the illicit cocaine scene in Harlem. In 1973, author Marc Olden attempted to cover the same ground in his book, *Cocaine.* Unlike *Dealer,* however, Olden's book bears the stamp of being a sensationalist piece of literature that does little to help clarify the complex relationship between drug use and the urban situation. Although there is no doubt that some cocaine-related happenings in New York City are criminal, debased, and, by almost anyone's definition, undesirable, it is presumptuous of Olden to imply that these actions are a result of using cocaine or that all the unsavory elements described in his book are necessary by-products of an association with the drug. Olden seems unduly influenced by the suspicions and anecdotes of the federal law-enforcement agents who helped him assemble much of the data for this book. Woodley's presentation of urban cocaine use is, in a sense, no less shocking or disturbing than Olden's, but it is clearly more sensitive and useful in terms of its choice and handling of the ethnographic material.

In 1972, Edward M. Brecher and the editors of *Consumer Reports* contributed heavily to the public's understanding of psychoactive substances. In their book *Licit and Illicit Drugs,* they not only acquaint their readers with the fact that, despite popular notions, alcohol, tobacco, and coffee are potentially dangerous and addicting drugs of abuse, they also use this fact to underscore how such less damaging and nonaddicting substances as coca are illogically subject to harsh social disapproval and unusually strict legal sanctions.

Just as important is their clear-cut separation of coca and cocaine. These two different but related substances were discussed under two different but related chapter headings and, accordingly, the authors stressed the physiological and psychological dissimilarities. By doing this, *Consumer Reports* laid a contemporary foundation for the construction of more meaningful laws and created a public knowledge of these poorly understood drugs. In this respect, knowledge is power, and knowledge of a particular drug and what it can and cannot do is the first step toward an intelligent association with the substance.

The effort to present accurate and unbiased information on cocaine was continued in 1974–75 with the publication of two useful readers on the subject. In one of these, David Solomon and George Andrews edited a unique collection of coca and cocaine papers that ranged from a translation of Paola Mantegazza's 1859 *Coca Experiences* to Jerry Hopkins's insightful look at contemporary cocaine use. The other reader was Robert Byck's *Cocaine Papers,* a translated collection of Freud's writings about the drug. This book sheds light on an interesting and important chapter in the history of cocaine. Because it deals with many of the nuances associated with the drug's rediscovery in the 1880s, Byck's *Cocaine Papers* amounts to a valuable educational tool that can be used to help understand some of the complexities of cocaine rediscovery in the 1970s.

In 1975–76, two more important books were published on cocaine use, Richard Ashley's *Cocaine: Its History, Uses and Effects* and *Cocaine: A Drug and Its Social Evolution* by Lester Grinspoon and James Bakalar. In the first of these, Ashley does an excellent job of exposing many of the myths surrounding cocaine and seems to be on his way to presenting a good explanation of the drug's physiological and psychoactive effects. At this point, however, he commits some grave sins of omission—in his appendices Ashley develops a tone that downplays cocaine's possible dangers and he fails to recognize some important considerations. For instance, Ashley conducts his own cocaine experiment by interviewing and observing eighty-one users and makes some sweeping and possibly misleading statements on the basis of these observations. He reports that none of his cocaine users experienced any depression or marked craving for the drug when their supply ran out and says that reports of a strong compulsion to resume use is a statement that is "devoid of meaning."[6] He goes on to extrapolate that paranoia and hallucinations are "rare" and "almost never reported by modern users."[7] He adds that almost all pharmacologists, physicians, and drug experts are ignorant of the effects of cocaine. He insists that users are "those best suited" to discuss the drug's effects.[8]

Ashley may be correct when he reports that none of the cocaine users *he observed* experienced any depression or craving for the drug* and he is technically correct in saying that cocaine does not produce the awful physiological addiction of the opiates. After reading Ashley's book, however, one is left with the impression that the consequences of cocaine use are close to being nonexistent and that the user is completely free from any compulsion to resume using it. There are only implicit illusions to the important abuse potential of cocaine and the other overt ways in which the misuse of this drug can upset the individual and the society.

Ashley appears to have been victimized by his own research. After uncovering so many brutal examples of the ways in which cocaine use has been misrepresented to the public, there is a great temptation to be suspect of any damning pieces of anticocaine information and a tendency toward overenthusiasm concerning the drug's abilities and harmlessness. Byck's *Cocaine Papers* demonstrate that Dr. Freud suffered from this same syndrome after writing *Über Coca.*

Much of what is wrong with Ashley's book has been compensated for by the publication of Grinspoon and Bakalar's work. Contrary to Ashley's claim about the ignorance of physicians and drug experts, Grinspoon (who is both) and Bakalar present a well-researched and thought-out book that successfully balances the pros and cons of cocaine consumption. After surveying the history of coca and cocaine, as well as all the myths and half-truths that have surrounded these substances, they launch into a carefully worded discussion of the effects of using cocaine. They point out that while it is entirely possible to enjoy a moderate and benign relationship with this drug, there are also some serious consequences that often go hand in hand with abusing it. Grinspoon and Bakalar caution their readers that "the more spectacular consequences of cocaine abuse are not typical of the drug's effects as normally used any more than the phenomena associated with alcoholism are typical of the ordinary consumption of that drug."[9] They then go on to document some horrifying tales of abuse that come from interviews with users—the same group that Ashley advises us to consult for accurate accounts of the drug's effect.

Through Grinspoon and Bakalar we hear stories of tactile, auditory, and, in some rare cases, visual hallucinations, paranoid delusions

*In pseudoscientific, do-it-yourself experiments of this nature, it is always difficult to determine the conditions under which this observation was made. For instance, did Ashley have an idea of the mental state of the user *before* that person took cocaine and was he with the user some hours or even a full day after the cocaine episode, a time when many users report the onset of depression or irritability.

that sometimes lead to unacceptable behavior,* irritable depression, and "the same kind of physiologically unspecific feeling of need that is associated with nicotine, caffeine, and amphetamines."[10] In other words, all the things that Ashley said we needn't worry about.

*One story involved a heavy cocaine sniffer who attacked his bathroom laundry basket with a hatchet thinking there was a tiny policeman hiding inside.

# 15

# Cocaine Magic, Cocaine Myth: A Recurring Historical Dilemma

This modern debate over the effects and consequences of cocaine use invites a general discussion on the topic of drug abuse, popular attitudes, and mythological beliefs.

It is well known that a drug user's reaction to any given psychoactive substance is governed by three important considerations: the first and most obvious of these is the established pharmacological action of the drug or the effects that are accepted by and listed in textbooks. The other factors are the set (the individual's expectation of what the drug will do to him) and setting (defined as the total physical and social environment in which the drug is taken).[1] Although these last two considerations are frequently overlooked, they are in fact so important that they can sometimes combine to overpower the drug's reported pharmacological actions. An example of this effect was presented earlier when Dr. Fronmuller's cocaine experiments were discussed. In those experiments, one of his patients was given .33 grams of pure cocaine (listed everywhere but in the U.S. legal documents as a CNS stimulant) and the patient reacted to that dose by falling asleep.

Other important considerations about set and setting and how these factors may affect the cocaine user come from Dr. Norman Zineberg, associate professor of clinical psychology at Harvard. Zinburg states that "the vaguer and less predictable the pharmacological effects of the drug . . . the greater the influence of set and setting."[2] Given cocaine's famous unpredictability* and accepting Zinberg's appealing logic, it is fair to assume that the set and setting are especially important factors in the understanding of how individuals will react to cocaine use in a particular situation. Furthermore, Zinburg reminds us that part of the set and setting of nonmedical drug use is the prevailing public attitude toward any particular drug at the time it is being used.

These popular attitudes give rise to and in turn are reinforced by the invention of mythologies. Like other aspects of human folklore, myths tend to develop and spread during anxious times when societies are in need of solutions to complex problems or when individuals seek justification for positions formed in defiance of fact.

At this point, it should be obvious how important public attitude and its subsequent mythologies have been in the history of coca and cocaine. Early coca myths deified the leaf and created a setting in which it was thought of as being divine. If the leaf was not actually worshiped in ancient Peru, it was surely a focal point of the culture. Reports of great harm to the Inca or any pre-Inca society because of coca use are nonexistent. From what is known about these

*Historically, cocaine had had a remarkable ability to deceive most of the Westerners who have come in contact with it.

societies and their relationship with the leaf, it seems that the only associated negative aspects were the political schemings and warfare that sometimes took place when interested parties competed for control of it and later the airs of exclusiveness that came to surround the leaf during the reign of Pachacuti and his sons.

Because of this exclusiveness, the coca leaf was transformed into an object of public enchantment that was actively, or in some cases frantically, pursued by Incas desiring status or tangible acclaim from their god/king. This glamorization process set the stage for the first reports of widespread Indian coca abuse in Peru. When Spanish conquistadors took control of the country and its resources, they further disrupted the coca sanctions by opening the doors to the government's warehouses and letting the much-sought-after leaf circulate freely among the masses. This was the equivalent of leaving a supermarket door unlocked during a famine.

The unsuspecting Spanish reacted in an understandable way to what they saw; they interpreted the Indian frenzy to get the leaf as acts of "vicious gluttony" and wretched excess that were most likely caused by the plant itself. Spanish priests, in turn, inaugurated a full-scale attack on the leaf that was heavy with religious incriminations. It was termed "a demonic manifestation from hell" and there was a call for complete and total eradication. Needless to say, the anticoca mythology of the Spanish clergy bears a striking resemblance to that adopted in the U.S. by prohibitionists Hamilton Wright and Harry J. Anslinger more than 450 years later.

Independent of the Spanish feelings about coca in Peru were their feelings about the plant when it arrived in the court of Charles V. In the face of religious inquisition and traditional models of courtly sophistication, the coca leaf did not even attract a mythology—it was ignored by a public living in one of the most unreceptive settings imaginable. This condition survived in Europe until Niemann extracted the cocaine alkaloid from the coca leaf and Sigmund Freud rediscovered both substances shortly thereafter. At that time cocaine became a "wonder drug" and a "magical substance" for Freud; his colleague, Carl Koller, quickly added to this positive set when he demonstrated its value as a "perfect local anesthetic." Soon after these enthusiastic announcements from Vienna, cocaine was favorably received all over Europe and America; but in the optimistic and positive setting of the Gay Nineties, the beneficial aspects of cocaine were exhalted to the point that some serious negative aspects of the drug were neglected or completely overlooked. In this setting, a procaine mythology developed and upset the necessary balance of pros and cons that must be maintained for an individual or society to enter into a successful relationship with a potentially dangerous drug. To compound the situation, the effects of coca and cocaine, and indeed even the terms themselves, became hopelessly confused. This confusion led to some episodes of accidental misuse through ignorance and added to the growing number of cocaine abusers.

The positive, enthusiastic, and uneducated public attitude toward cocaine produced a noticeable number of abusers as well as an anxious number of nondrug-using U.S. citizens who feared the worst. They declared war on the social setting that tolerated and even endorsed the use of drugs and they conducted their campaign against cocaine use with the blind zeal of religious crusaders. Their weapons were propaganda attacks designed to convince the nation of an intrinsic evil that necessarily lurks in every grain of cocaine. Their strategy was as extreme as it was effective. Once the reader (or lawmaker) was persuaded that cocaine was bad in itself, there was no longer any need to try to explain the effects of different doses, or to separate coca use from cocaine use, or to differentiate the user from the abuser. No matter what the variable, cocaine was considered bad stuff and anyone who associated themselves with it was thought of as being equally malevolent.

This simpleminded and easily digestible dogma became popular and powerful enough to counteract the procaine mythology that had its roots in Vienna and the Gay Nineties. Once again, public attitude toward cocaine had taken a radical and negative shift and had created a correspondingly negative mythology; in the midst of the propaganda assault that helped create this unfavorable setting federal laws were enacted that, by their wording and implication, added to the problem and created their own special brand of negativity.

The laws against cocaine, alcohol, and other drugs aggravated an already bad situation and made it worse. They enriched the newly forbidden substances with cheap status, drove drug users underground, invited organized crime, and created outrageously high prices and adulteration; they brought the law into a state of disrepute and, most important of all, they did not work! In the end, these ill-conceived and untimely prohibitions created possibly the most negative setting of all—an anticocaine mythology that was so strong and so well publicized and perpetuated by law that it would still bear an unfavorable effect on users and society more than sixty years later.

The cocaine situation in the United States today is in many ways very similar to the situation that existed here and in Europe in the 1920s. Illicit, expensive, adulterated, and status-enriched cocaine is conspicuously apparent in all strata of the society. As in the 1920s, more and more police and law-enforcement agents are assigned to halt the cocaine traffic and the laws are continually stiffened, but these efforts prove meaningless and the drug's popularity and presence continues to grow. Because of the setting created by the drug laws, users in both the 1920s, the 1970s and the 1980s find themselves in disreputable or outright criminal situations. The few who can easily afford the 100-dollar street gram complain that they tend to use the drug too much, and they regret it.* Those who cannot afford it sometimes end up performing dangerously unfamiliar acts to get it. A good example of this is the person who discovers that he can deal in a quantity of the drug to get a little free. Suddenly, this person (who only wants a little cocaine to use) finds himself making cash-for-drugs exchanges in a lawless social environment filled with undercover agents, thieves, and violence.

Probably the worst effect of the drug laws, however, is their capacity to transform simple plants or laboratory alkaloids into high-priced luxury items. When a drug or any other consumable item falls into this category, there is an accompanying tendency toward excess. When a drug such as cocaine (whose potential for abuse through excess is already very high) is placed in this category, the results can be disastrous.* A whole syndrome of events begins to grow up around the drug in this situation and they directly affect the cocaine users' reaction. For instance, in the 1970s as well as the 1920s, a forbidden-fruit aura surrounded cocaine. Its illicit status and unaffordability gave it a kind of romantic irresistibility. There is a desire to use more and more of it not only because it makes one feel good, but also because it is considered a luxury. When this latter factor is coupled with cocaine's pharmacological ability to deliver a short period of happy excitement that can be quickly duplicated with another small sniff, the result is a unique psychological and physiological invitation for abuse that amounts to something that could be called cocaine addiction.

The desire to use excessive amounts of the cocaine alkaloid repeatedly over a long period of time (or, in other words, abuse it) is not simply a physiological matter; if it were, cocaine could be called an addictive drug and the hospitals would be crowded with patients suffering from cocaine withdrawal symptoms. But cocaine is not a truly addicting drug; in fact, there are reports of cocaine repulsion after large doses (William Alexander Hammond) or prolonged use (Sigmund Freud). While the abuse potential of cocaine is always pharmacologically present, this potential seems to be significantly aggravated, and in some cases even triggered by, the psychological effects of both positive and negative settings and the existence of their corresponding mythologies.

Pachacuti manipulated the leaf and created a mythology around it that made it an exclusive prize. In this setting, the abuse potential of coca (small as it is) was aggravated and there was evidence of excessive indulgence. When the Spanish entered the picture, they recognized the strength of the coca mythology and the excess it caused, and fought to alter the situation and end coca use by the creation of their own myths.

The Spanish myths about coca were current until Freud and Koller rearranged the setting and unknowingly added to the creation of a strong procaine mythology that would endorse use to the point of abuse in the 1890s. Under the impression given by the greatest purveyors of the myth, the patent-medicine manufacturers, that cocaine was a harmless cure-all, restorative, and recreational device, many turn-of-the-century users became abusers and their actions began to attract some staunch critics. These cocaine critics, like the Spanish before them, were willing to go to almost any length to scuttle the favorable reputation that cocaine had established and demonstrate that the use of this frightful substance would inevitably lead to disaster and ultimate ruination. The reactionary anticocaine mythology that developed in the 1920s and 1930s gave rise to the drug laws and a national fear and hatred of the drug itself.

*A new menace called "freebase" is causing a tremendous amount of trouble and regret among those unlucky enough to become involved with it. Freebase is a further concentration of the already concentrated alkaloid and lends support to the observation that as far as using cocaine is concerned, the farther away from the leaf you get, the more trouble you're in. A freebase habit can cost $1000 a day. With freebase, the strongest are not strong enough.

*Evidence the self-styled cocaine supermen of the 1980s who are methodicaly self-destructing at the alarming rate of a gram a day.

The unfortunate few who pursued their cocaine habits into the 1920s and 1930s often fell victim to one of the pitfalls produced by this strong antidrug mythology. Some believed that cocaine did nothing more than lift the spirits and numb the nose. If these people could afford cocaine, they often ended up using it continually until it was gone, and then began over again until their cash reserves, mucous membranes and personalities gave out. Those who still wanted to use cocaine but couldn't afford it after the drug laws were passed found themselves engaged in a high-priced criminal adventure entailing a variety of negative consequences. Other people knew that cocaine was dangerous in excess, but the drug's fashionable allure lent psychological impetus to its physiological abuse potential, and the resulting combination produced a package that was so attractive that abuse was almost inevitable. Essentially, the cocaine pitfalls of the 1920s were caused by a procaine mythology (a belief it was harmless) or else a psychological abuse factor (status enrichment) produced by the particular social setting.

For the next thirty years, the nonmedical use of cocaine was little more than a memory. However, during this same time, the stringent anticocaine mythology of the 1920s was not only maintained, but actually fortified by anxious law-enforcement agents and lawmakers who had been indoctrinated with an outdated philosophy.

The perpetuation of the anticocaine mythology of the 1920s hastened the creation and passage of even tighter drug-control laws and had a significant effect on the setting in which drugs such as cocaine were used. Because of the laws, cocaine was still an exclusive, high-priced luxury item in the 1970s just as it was in the 1920s; it continued to carry the psychological stigma that is so disastrous when combined with a drug that has the pharmaceutical properties of cocaine.

When drug use came in vogue again around 1965, interested individuals were faced with what they correctly regarded as unreasonable resistance from society at large. As has been the case throughout the history of coca and cocaine, opponents of the established mythology (who were in this case the drug users) moved to counteract the mood of the existing setting by creating their own myths. Traditionally, the tenets of the new mythology are expressed in inverse proportion to those of the old one.

As far as cocaine is concerned, the new mythology that developed had as much potential for disaster as the old one; the danger was that, while reacting to an obvious injustice, there was a tendency to overreact and go as far to the left as perhaps the old drug laws had gone to the right. With cocaine, this may have encouraged excessive consumption by those who were convinced that the drug was without physiological or psychological consequences. They were cheered on in this belief by contemporary publications that implied that such was the case.

In an attempt to bolster its authenticity, the new mythology recruited and misinterpreted the results of modern scientific investigations into the effects of cocaine. The pharmaceutical revelation that cocaine is nonaddictive, non-narcotic, and without tolerance or withdrawal symptoms has somehow amounted to a tacit medical approval of the drug in the minds of those converted to the new mythology. This is a slightly modified version of one of the key reasons for the cocaine-abuse problem around the turn of the century. Because the drug laws are still present and operating in full force, the glamorization and status-enrichment aspects of cocaine use are also present now, and they represent the primary stimuli for the cocaine abuse that took place just after the Spanish conquest and also in the 1920s, just after the laws were first passed.

So in the 1970s, under the new mythology, all the classic ingredients that prompted cocaine abuse in the past were alive and at work at one time.

In a 1975 article entitled "Cocaine: Current Rage for the Rich and Hip," Nicholas Von Hoffman called cocaine "the most social of drugs" and perceptively observes, "When a drug becomes the rage, people won't listen to anybody's warnings against it."[3] Jerry Hopkins had the same thing in mind when he said:

> Why would anyone pay up to 10,000 percent mark-up on a product he knows has been brutally adulterated? In the glittery, shattered early 70's, cocaine offered more than chemical euphoria—it also offered a higher station in life. . . . In a drug-oriented society, coke had become the gourmet trip.[4]

These warnings were based on an implicit understanding of the new cocaine mythology and the realization that it was being practically applied throughout the United States. In 1977, *Newsweek* magazine featured a long article on cocaine, declaring it to be "the recreational drug of choice for countless Americans," and supported this claim by saying:

> Businessmen use it to get going in the morning and entertainers use it to keep going at

> night. College students and housewives use it, stockbrokers and fashion designers use it, rock singers and used-car salesmen use it.[5]

This list could be expanded. Government officials agree that the traffic in illegal cocaine is presently more than a one-billion-dollar a year business in the United States and that there is no end in sight. In other words, the reality and proportion of contemporary cocaine use is a subject that is not open to much debate.

That cocaine is, and has been since its creation, a recurring and significantly influential force in American culture makes the issue of how to relate to this substance a very important one. In order to establish a setting that correctly balances the pros and cons of cocaine use, the practice of inventing mythologies to support or attack the drug must be completely abandoned. It is essential that individuals and government first evaluate the drug as honestly and intelligently as possible and then, from this educated position, develop more moderate and benign associations with the drug or, if needed, institute legislative control measures that are aimed at those members of the society who refuse to do so.

# 16

# Epilogue

At this writing there is not a great deal known about the clinical effects of chronic but moderate cocaine use; indications so far tend to suggest that the process could be a relatively harmless one. There is, however, an overwhelming amount of data confirming that cocaine abuse is responsible for a great deal of human misery and suffering. From historical survey it is also known that the desire to use cocaine is given to periods of intense popular florescence that may occur in any society at any time and in spite of any law that says otherwise. This is generally the way with human behavior and drug use.

With this information in mind, it would seem important to pursue almost any reasonable procedures that would work to minimize the abuse of cocaine when a society becomes attached to it. As I have already mentioned, the dispelling of the different cocaine mythologies would be a good first step in this direction. A second recommendation involves the formulation of laws that punished abusers and ignored users who cause no harm to the community. While I believe that these two suggestions would go a long way toward the solution of this country's cocaine problems, there is still a final consideration that is worthy of some attention.

If cocaine proves to be harmless in long-term moderation (as it seems to be for South American Indians and other little-publicized moderate alkaloid users), then we need only concern ourselves with eliminating the conditions that inspire excessive use and invite disaster.

One of the most obvious yet effective ways of combatting cocaine abuse while at the same time guaranteeing the freedom of individual access to the drug is to make cocaine available everywhere in moderate oral doses that are nearly impossible to abuse. Such was the case with Mariani's wine and the pre-1906 version of Coca-Cola. With the former, an alcoholic collapse would take place long before a significant amount of cocaine could be consumed; with the latter, tooth decay would be a more pressing concern than the amount of cocaine ingested. A modern tourist in Peru can swell his belly with coca tea and simultaneously swell his cheek with coca leaves and still not come close to having a dangerous amount of cocaine in his body. In the same vein, it would be difficult or impossible to imagine such prominent figures as Pope Leo XIII, the Prince of Wales, or Thomas Alva Edison locked into a frantic, desperate, or violent behavior pattern as a result of chronic but moderate oral doses of cocaine.

Recently, Dr. Joel Hanna of the University of Hawaii has presented evidence suggesting that this situation would be highly unlikely. Looking at coca use from a biocultural perspective, he points out that, first, the consumption of coca in the Andes "does not appear to have a serious effect on community life since the . . . characteristic[s] of cocaine addiction are not witnessed in the behavior of the coqueros."[1] He supports this observation by taking Montiesino's 1965 figure of 350 mg. as the maximum amount

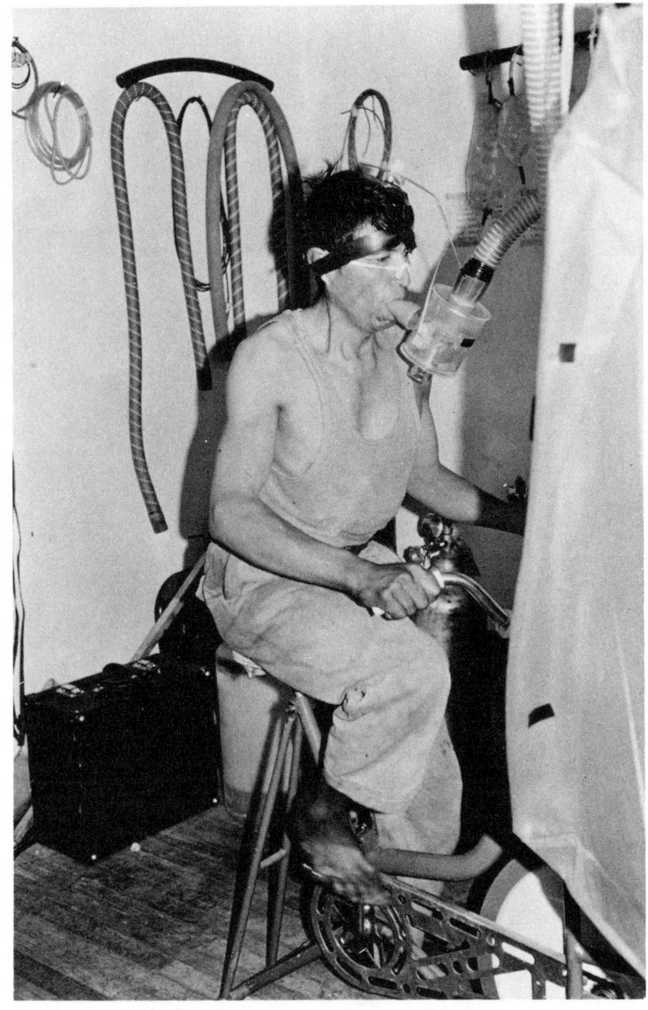

**Part of Dr. Hanna's coca experiments in southern Peru.** ***(Courtesy of Dr. Joel Hanna)***

of cocaine that could be extracted from 50 grams of leaves (broken down into three chews a day), and then points out that with these figures, less than 0.03 grams* of cocaine would make it into the liver (in humans, the principal site of cocaine detoxification) on the first pass.

Next, like Monge, Hanna concludes that coca use may be an adaptive force in an environment such as the Andean pleateau. His report demonstrates that coca warms the Indian, quells his hunger, reduces his fatigue, and is high in nutritional value.

Later on, like most responsible researchers, Hanna mentions that the physiological effects of coca and cocaine "should be unequivocally distinct."[2] He goes on to caution, however, that cocaine may not necessarily be totally responsible for the stimulation associated with coca use; the amounts of arecoline and the nicotine that are contained in the leaf may also be significantly active factors. Hanna's general suspicions are supported by Dr. Otto Nieschulz of the chemical firm, Promonta GmbH., who said, "Cocaine, ecgonine or cuskohygrine alone or together could cause the psychic effects of coca use, its influence on physical performance, and its antidepressive effects."[3]

So, to date, little is known about the other alkaloids in coca and how they relate to each other and to the effects of using the leaf. This gap in our knowledge about the coca alkaloids is a potentially significant one because millions of Coca-Cola drinkers around the world regularly consume these ingredients. Since the passage of the Pure Food and Drug Act in 1906, the Coca-Cola Company has gone to some trouble in its effort to remove all traces of the cocaine alkaloid from the coca leaves they use to flavor their popular confection, but, according to the information presented by Hanna and Nieschultz, they may have not succeeded in eliminating all the active ingredients.

The traces of nicotine, arecoline, ecgonine, and cuskohygrine, together with the sugar and caffeine the beverage contains, may very well justify the slogan that says drinking Coca-Cola is a "pause that refreshes."

From this angle, it can be said that we have on the world market today a relatively harmless beverage that contains stimulating and potentially dangerous properties from the plant world. The same thing can be said of coffee and tea. Needless to say, these products are almost universally accepted and available, legal, and even encouraged through prolific advertisements. Moreover, these substances have practical and social applications that have become traditions in many modern cultures.

The abuse of these mildly psychoactive beverages is generally considered to be a minor personal weakness or, at worst, a disgusting but harmless habit. Accordingly, there have been no international movements to uproot the world's coffee plants or to halt the production of Coca-Cola. Instead, society has developed informal sanctions that are appropriate to the situation. A heavy coffee drinker may acquire the reputation of being "speedy," jittery, or possessed of bad breath; likewise, a person who consumes ten bottles of Coca-Cola a day may suffer some good-natured reproach from friends who might playfully refer to his "Coca-Cola addiction." These

*Cocaine addicts sometimes use as much as 1 or 2 full grams a day against the mucous membranes or introduced directly into the veins; a far cry from the 0.03 grams orally ingested by Andean Indians.

social safety valves work to head off potentially dangerous scenes that involve the unchecked use of desirable substances, or the creation of myths and laws that grossly misrepresent the effects of normal use.

To reach this kind of understanding with cocaine, it appears necessary to keep the alkaloid contained within the leaf as it naturally occurs and, like caffeine, use it orally and moderately to achieve the desired effect. With these provisions, it would be difficult to make a moral or legal separation between sipping a cup of coffee in New York City and chewing a handful of coca leaves in Cuzco.

# Notes

## Chapter 1. An Ancient Beginning

1. Richard T. Martin, cited in George Andrews and David Solomon, *The Coca Leaf and Cocaine Papers* (New York and London: Harcourt Brace Jovanovich, 1975), p. 22.
2. Alden J. Mason, *The Ancient Civilization of Peru* (Baltimore, Md.: Penguin Books, 1957), p. 143.
3. Howard Osborne, *South American Mythology* (London: Hamlyn Publishing Group, 1968), p. 238.

## Chapter 2. The Incas

1. Burr Cartwright Brundage, *Empire of the Incas* (Norman, Okla.: University of Oklahoma Press, 1963), p. 13.
2. Ibid., p. 47.
3. Ibid.
4. Osborne, *South American Mythology*, p. 89.
5. Anonymous, *Gourmet Cokebook* (Washington, D.C.: Whitemountain Press, 1972).
6. Simone and Roger Waisbard, *Masks, Mummies and Magicians* (London: Oliver and Boyd, 1965), p. 89.
7. A. A. Moll, *Aesculapius in Latin America* (Philadelphia: W. B. Saunders Co., 1944), p. 13.
8. William Mortimer, *History of Coca* (San Francisco, Calif.: And/Or Press, 1974), p. 17.
9. Garcilaso de la Vega. *El Inca: Royal Commentaries of the Incas* (Austin, Tex.: University of Texas Press, 1966), p. 330.
10. Bertrand Flornoy, *World of the Incas* (New York: Vanguard Press, 1956), p. 135.
11. Ibid.
12. Hans Baumann, *Gold and Gods in Peru* (London: Oxford University Press, 1963), p. 40.
13. Flornoy, *World of the Incas*, p. 162.
14. Louis Baudin, *Daily Life in Ancient Peru.* (London: Allen & Unwin, Ltd. Museum St., 1961), p. 41.
15. Brundage, *Empire of the Incas*, p. 111.
16. Carl O. Sauer, "Cultivated Plants of South America." In *Handbook of South American Indians.* Julian Stewart, ed. Bureau of American Ethnology. Washington, D.C.
17. Margret Towle, *Ethnobotany of Pre-Columbian Peru* (New York: Viking Fund Publications in Anthropology no. 30, 1961), p. 21.

## Chapter 3. The Most Valuable Substance

1. Brundage, *Empire of the Incas*, p. 98.
2. Ibid., p. 100.
3. Mortimer, *History of Coca*, p. 36.
4. Ibid., p. 37.
5. Brundage, *Empire of the Incas*, p. 106.
6. Mortimer, *History of Coca*, p. 233.
7. Ibid., p. 234.
8. G. H. S. Bushnell, *Peru* (New York and Washington: Frederick A. Praeger, 1957), p. 136.
9. José de Acosta, *The Natural History of the Indians* (London: Hakluyt Society, 1880), p. 245.
10. Louis Baudin, *Daily Life in Ancient Peru*, p. 88.
11. Brundage, *Empire of the Incas*, p. 47.
12. Mortimer, *History of Coca*, p. 20.
13. Ibid., p. 154.
14. Ibid., p. 47.
15. Ibid., p. 48.
16. Ibid., pp. 68, 46.
17. Ibid., p. 72.
18. Waisbard, *Masks, Mummies and Magicians*, p. 62.
19. Loren MacIntyre, *The Incredible Incas and Their Timeless Land* (Washington, D.C.: National Geographic Society, 1974), p. 74.
20. Erwin H. Ackerknecht, *Medical Practices Handbook of South American Indians*, Julian Stewart, ed. (Washington, D.C.: Bureau of American Ethnology, 1949), p. 638.
21. R. L. Moody, "Studies in Paoleopathology. Injuries to the Head among the Pre-Columbian Peruvians," *Annals of Medicinal History* 1 (1927): 237.
22. Biejer-Priesto, H. M.D. "Coca Leaf and Cocaine Addiction: Some Historical Notes." *Canadian Medical Journal* 93 Sept. 25, 1965, p. 703.

## Chapter 4. A Small Band of Curious Men

1. Brundage, *Empire of the Incas*, p. 224.
2. Sacheverell Sitwell, *Golden Wall and Mirador* (Cleveland: World Publishing Co., 1961), p. 117.
3. Edward Hymans and George Ordish, *Last of the Incas.* New York: Simon and Schuster, 1963.
4. Brundage, *Empire of the Incas*, p. 243.
5. Ibid., p. 224.

6. Sigmund Freud, cited in Dr. Robert Byck, *The Cocaine Papers* (New York: Stonehill Press, 1974), p. 50.

7. Amerigo Vespucci, cited in Francisco Guerra, *The Pre-Columbian Mind* (London: Seminar Press, 1971), p. 47.

8. Brundage, *Empire of the Incas,* p. 303.

9. Q.M. Stephen-Hassard, "Sacred Plant of the Incas," in *Pacific Discovery* 23, no. 5 (September–October 1970): 28.

### Chapter 5. Imperial Spain and the Inca Leaf

1. Quoted in John Hemmings, *Conquest of the Incas* (New York: Harcourt Brace Jovanovich, 1970), p. 120.

2. Gordon Sharp, "Coca and Cocaine Studied Historically," in *Pharmaceutical Journal* 82, (1909), p. 29.

3. Hemmings, *Conquest of the Incas,* p. 365.

4. Quoted in Mortimer, *History of Coca,* p. 108. p. 139.

5. Arthur F. Zimmerman, *The Reign of Viceroy Toledo Vth, Viceroy of Peru* (Caldwell, Idaho: Claxton Printers Ltd., 1938).

6. Pedro Ciezea de Leon, *The Incas of Pedro de Ciezea de Leon,* Von Hagen, ed. (Norman, Okla.: University of Oklahoma Press, 1959), p. 352.

7. Mortimer, *History of Coca,* p. 160.

8. Hemmings, *Conquest of the Incas,* p. 368.

9. Biejer-Priesto, "Coca and Cocaine—Historical Notes", p. 781.

10. Zimmerman, *Reign of Viceroy Toledo,* pp. 639–41.

11. Domingo de Santo Tomas, quoted in Hemmings, *Conquest of the Incas,* p. 370.

12. Acosta, *Natural History of the Indians,* p. 245.

13. Lillian E. Fisher, *The Last Inca Revolt, 1780–1783* (Norman, Okla.: University of Oklahoma Press, 1966), p. 79.

### Chapter 6. Pirates, Priests, and Poets

1. Mortimer, *History of Coca,* p. 154.

2. Hemmings, *Conquest of the Incas,* p. 309.

3. Francisco Hernandez, quoted in Guerra, *The Pre-Columbian Mind,* p. 137.

4. Bernabe Cobo, quoted in Mortimer, *History of Coca,* p. 192.

5. Sharp, "Coca and Cocaine," p. 28.

6. Abraham Cowley, quoted in Mortimer, *History of Coca,* pp. 26, 27.

7. Mortimer, *History of Coca,* p. 230.

8. Phillip A. Means, *Fall of the Inca Empire* (New York and London: Charles Scribner's Sons, 1935), p. 223.

9. Carl O. Sauer, *Early Spanish Main* (Berkeley, Calif.: University of California Press, 1966), p. 115.

10. Ibid., p. 135.

11. John M. Cooper, "Stimulants and Narcotics," in *Handbook of South American Indians,* Julian Stewart, ed. (Washington, D.C.: Bureau of American Ethnology, 1949), p. 549.

12. John Esquemelling, *Buccaneers of America* (London, George Routledge and Sons, 1684–85), p. 406.

13. Atkinson, *Magic, Myth and Medicine,* p. 95.

14. *Bulletin on Narcotics,* "Commission of Enquiry on the Coca Leaf," No. 1 (Oct. 1952), p. 23.

15. H. G. Haile, *The History of Johann Faustus* (Urbana, Ill.: University of Illinois Press, 1965), p. 8.

16. René Fulop-Miller, *Triumph Over Pain* (New York: Bobbs-Merrill, 1938), p. 14.

17. Norman Grover, "Man and Plants against Pain," in *Economic Botany* 19 (1965): 105.

18. Blas Valera, quoted in Garcilaso, *El Inca Royal,* p. 509.

19. Riennard Federmann, *The Royal Art of Alchemy* (Philadelphia: Chilton Book Co., 1964), p. 131.

### Chapter 7. The Age of Enlightenment

1. Means, *Fall of the Inca Empire,* p. 253.

2. Anonymous, *Gourmet Cokebook,* p. 13.

3. Mortimer, *History of Coca,* p. 230.

4. James Burney; *A Chronological History in the South Seas and Pacific Ocean* (London: Kegan Paul Ltd., 1817), 4:487–500.

5. Mortimer, *History of Coca,* p. 230.

6. Edward J. Goodman, *Explorers of South America* (New York: Macmillan, 1972), p. 185.

7. Ibid., p. 189.

8. Ulloa, quoted in Mortimer, *History of Coca,* p. 167.

9. Charles Darwin, quoted in Von Hagen, *South America Called Them* (New York: A. Knopf, 1945), p. 168.

10. Baron von Humboldt, *Personal Narrative of Travels of the Equinoctial Regions of the New Continent During the Years 1799–1814* (London: Kegan Paul Ltd., 1814), 1:41.

11. Ibid.

12. Van Hagen, *South America Called Them,* p. 4.

13. Editorial, *Gentlemen's Magazine* 12 (September 1814), p. 22.

14. Johan von Tschudi, quoted in Mortimer, *History of Coca,* p. 171.

15. Dr. Paolo Mantegazza, quoted in Mortimer, *History of Coca,* p. 137.

16. George Andrews and David Solomon, *The Coca Leaf and Cocaine Papers* (New York and London: Harcourt Brace Jovanovich, 1975), p. 38.

17. Karl Scherzer, *Voyage on the Novara. Narrative of the Circumnavigation of the Globe by the Austrian Frigate Novara* (London: Saunders, Otley and Co., 1861–63), p. 403.

18. Ibid.

19. Albert Niemann, quoted in Mortimer, *History of Coca,* p. 297.

20. Robert Christison, quoted in Mortimer, *History of Coca,* p. 366.

21. Andrew Weil, "A Gourmet Coca Taster's Tour of Peru," in *High Times,* May 1976, p. 334.

22. G. F. Dowdeswell, quoted in Mortimer, *History of Coca,* p. 426.

23. Ibid.

24. Angelo Mariani, *Coca and Its Therapeutic Action,* trans. J. N. Jarnos (New York: privately printed by A. Mariani, 1890), p. 10.

### Chapter 8. Freud, Koller, and Local Anesthesia

1. Ibid., pp. 14, 15.

2. Edward Kremers and Richard Udang, *History of Pharmacy* (Philadelphia: Lippincott, 1963), p. 157.

3. C. S. Rafinesque, quoted in Kremers and Udang, *History of Pharmacy,* p. 160.

4. Becker, cited in Byck, *The Cocaine Papers,* p. 275.

5. G. A. Ward, *Medical Record* 17 (1880): 497.

6. W. H. Bentley, quoted in Byck, *The Cocaine Papers,* p. 17.

7. J. H. Woods and D. A. Downs, "The Psychopharmacology of Cocaine," in *Drug Use in America: Problem in Perspective, The Technical Papers of the Second Report of the National Commission on Marijuana and Drug Abuse* (Washington, D.C.: U.S. Government Printing Office, 1973), 1:116.

8. Joel M. Hanna, "Coca Leaf Use in Southern Peru: Some Biosocial Aspects," in *American Anthropologist* vol. 34 (1974): 3.

9. *Louisville Medical News,* quoted in Byck, *The Cocaine Papers,* p. 21.

10. Sigmund Freud, quoted in Byck, *The Cocaine Papers,* p. 62.

11. Theodor Aschenbrandt, quoted in Byck, *The Cocaine Papers,* pp. 21–23.

12. Ibid., p. 23.

13. Ibid., p. 25.

14. Byck, *The Cocaine Papers,* p. 26.
15. Sigmund Freud, quoted in Byck, *The Cocaine Papers,* p. 264.
16. Ibid., p. 6.
17. Ibid., p. 67.
18. Ernest Jones, in Byck, *The Cocaine Papers,* p. 81.
19. Becker, cited in Byck, *The Cocaine Papers,* p. 274.
20. Sigmund Freud, cited in Byck, *The Cocaine Papers,* p. 7.
21. Ibid., p. 11.
22. Becker, cited in Byck, *The Cocaine Papers,* p. 276.
23. Carl Koller, cited in Byck, *The Cocaine Papers,* p. 283.
24. Ibid.
25. Sigmund Freud, cited in Byck, *The Cocaine Papers,* pp. 42, 43.
26. Carl Koller, cited in Byck, *The Cocaine Papers,* p. 247.
27. Ibid.
28. Ibid., p. 283.
29. Gaertner, cited in Byck, *The Cocaine Papers,* p. 285.
30. G. Liljestrand, "History of Local Anesthesia," in *International Encyclopedia of Phramacology and Therapeutics* (Elmsford, N.Y.: Pergamon Press, 1911), p. 19.
31. Sigmund Freud, cited in Byck, *The Cocaine Papers,* p. 294.
32. Carl Koller, cited in Liljestrand, "History of Local Anesthesia," p. 19.
33. Ibid.
34. Ibid.
35. Ibid.
36. Liljestrand, "History of Local Anesthesia," p. 23.

## Chapter 9. The First Cocaine Disasters

1. H. H. Rusby, *Jungle Memories* (New York: McGraw-Hill Book Co., 1933), p. 3.
2. Ibid., p. 100.
3. Sigmund Freud, cited in Byck, *The Cocaine Papers,* p. 117.
4. Leonard Corning, cited in Geoffrey Marks and W. K. Beatty, *Medical Gardens* (New York: Charles Scribner's Sons, 1971), p. 49.
5. David Musto, cited in Byck, *The Cocaine Papers,* p. 369.
6. William A. Hammond, cited in Byck, *The Cocaine Papers,* pp. 46–49.
7. Ibid., p. 184.
8. Sigmund Freud, cited in Byck, *The Cocaine Papers,* p. 173.
9. *National Clearinghouse for Drug Abuse Information,* Series 11 (Washington, D.C.: U.S. Government Printing Office, January
10. Woods and Downs, "Psychopharmacology of Cocaine," pp. 128–29.
1972), p. 6.
11. Nancy Elsworth, David Smith et al., "Current Perspectives on Cocaine Use in America," in *Journal of Psychedelic Drugs* 5, no. 2 (Winter 1972): 156.
12. Edward M. Breecher, *Licit and Illicit Drugs.* The Consumers' Union Report on Narcotics, Stimulants, Depressants, Inhalants, Hallucinogens and Marijuana—Including Caffeine, Nicotine and Alcohol (Boston-Toronto: Little, Brown and Co. Consumers' Union, 1972), p. 275.
13. Ibid.
14. G. A. Deneau et al., "Self Administration of Psychoactive Substances by the Monkey," in *Psychopharmacologia* 16 (1969): 227.
15. Bernfield, cited in Byck, *The Cocaine Papers,* p. 342.
16. Sigmund Freud, cited in Byck, *The Cocaine Papers,* p. 121.
17. Ibid., p. 172.
18. Ibid., p. 174.
19. Jones, cited in Byck, *The Cocaine Papers,* p. 199.
20. Ibid., p. 200.
21. Sigmund Freud, cited in Byck, *The Cocaine Papers,* p. 205.
22. G. R. Gay, et al., "Cocaine in Perspecitve: Gift from the Sun God to Rich Man's Drug," in *Drug Forum* 2, no. 4 (Summer 1973): 414.

## Chapter 10. Belle Epoque

1. Myron G. Schultz, "The 'Strange Case' of Robert Louis Stevenson," *Journal of American Medical Association* 216, no. 1 (5 April 1971): 90.
2. Fanny Stevenson, cited in Schultz, "The 'Strange Case' of R. L. Stevenson," p. 91.
3. Arthur Conan Doyle, *The Sign of Four* (New York: Crown Publishers, 1967), p. 1.
4. Ibid., pp. 1–2.
5. Mortimer, *History of Coca,* p. 178.
6. E. J. Kahn, Jr., *The Big Drink: The Story of Coca-Cola* (New York: Random House, 1950), p. 157.

## Chapter 11. Patent Medicines and Drug Laws

1. *Nostrums and Quackeries*-Special publication of the *Journal of the American Medical Association.* Chicago 1912, p. 427.
2. Woods and Downs, "Psychopharmacology of Cocaine," p. 118.
3. *National Clearinghouse,* p.5.
4. Ibid., p.4.
5. Edward H. Williams, "The Drug Menace in the South," in *Medical Record* 85 (1914): 249.
6. Ibid., p. 247.
7. Anonymous, *"The Cocaine Habit," The Military Surgeon,* 4, 1914.
8. Christopher Koch, *Literary Digest,* "Cocaine," 28 March 1914, p. 687.
9. Annie C. Meyers, quoted in Cleveland Moffet, "Eight Years in a Cocaine Hell," *Hampton's Magazine* 24, no. 1 (1911): 25.
10. Moffett, "Cocaine Hell," p.27.
11. New York *Times,* 24 June 1914, p. 18.
12. Ibid., 18 December 1918, p. 34.
13. Williams, "Drug Menace in the South," p. 247.
14. Ibid.
15. Ibid., p. 248.
16. J. W. Watson, "The Cocaine Problem," New York *Tribune,* 21 June 1903, p. 52.
17. Dr. Hamilton Wright, quoted in David F. Musto, *The American Disease—The Origins of Narcotic Control* (New Haven, Conn.: Yale University Press, 1973), pp. 44–45.
18. Christopher Koch, quoted in Musto, *The American Disease,* p. 647.
19. Musto, *The American Disease,* p. 7.
20. Charles B. D. Towns, "The Peril of the Drug Habit," *Century Magazine,* 1912, p. 586.
21. Cleveland Moffet "Cocaine Hell," *Hamptons Magazine,* p. 27.
22. Samuel Hopkins Adams, "Great American Fraud," in *Colliers Magazine,* 8 June 1907, p. 155.
23. Angelo Mariani, cited in Byck, *The Cocaine Papers,* p. xxxviii.
24. Charles L. Mitchell, cited in *Nostrums and Quackeries,* p. 365.
25. Gerald T. McLaughlin, "Cocaine: The History and Regulation of a Dangerous Drug," in *Cornell Law Review* 58, no. 3 (March 1973): 568.
26. Ibid.
27. Francis Burton Harrison, cited in Musto, *The American Disease,* p. 46.
28. Hamilton Wright, cited in Rufus King, *The Drug Hangup—America's Fifty Years Folly* (New York: W. W. Norton, 1972), p. 21.
29. Ibid., p. 33.
30. Lester Volk, cited in King, *The Drug Hangup,* p. 21.
31. Musto, *The American Disease,* p. 4.

### Chapter 12. Illicit Cocaine in the 1920s

1. Charles E. Goshen, *Drinks, Drugs and Do-Gooders* (New York: The Free Press, 1973), p. 21.
2. King, *The Drug Hangup*, p. 30.
3. Al Smith, cited in King, *The Drug Hangup*, p. 30.
4. Andrews and Solomon, *The Coca Leaf and Cocaine Papers*, p. 5.
5. Norman E. Zinberg and John A. Robertson, *Drugs and the Public* (New York: Simon and Schuster, 1972), p. 77.
6. Goshen, *Drinks, Drugs and Do-Gooders*, p. 5.
7. Joseph Gagliano, "A Social History of Coca in Peru" (Diss., Georgetown University), Washington, D.C., 1960).
8. I. C. Chopra and Sir R. N. Chopra, "The Cocaine Problems in India," in *Bulletin on Narcotics*, vol. 10, no. 2, April–June 1958, p. 13.
9. Ibid.
10. Dr. Nathan Mutch, "Cocaine," in *Guys Hospital Gazette* 46 (1932): 425.
11. *New York Times*, 16 August 1925, p. 41.
12. Magnus Hirschfield, *Curious Sex Customs in the Far East* (New York: Grosset and Dunlap, 1935), p. 238.
13. Dino Segre (Pitigrilli), *Cocaine* (San Francisco: And/Or Press, 1974), p. 52.
14. Ibid., p. 65.
15. Marcel Proust, *The Past Recaptured (Á la Recherche du Temps Perdu)*, trans. (London: Chatto and Windus, 1952), p. 283.
16. Aleister Crowley, *Diary of a Drug Fiend* (New York: Samuel Weiser, 1973), p. 244.
17. Herman Hesse, *Steppenwolf* trans. (New York: Holt, Rinehart and Winston, 1964), p. 141.
18. Ibid., p. 150.

### Chapter 13. The Movement to Exterminate Coca

1. Andrews and Solomon, *The Coca Leaf and Cocaine Papers*, p. 6.
2. Fernando Cabieses Molina, cited in Andrews and Solomon, p. 265.
3. United Nations Economic and Social Council, *Commission on Narcotic Drugs, 6th Session, Lake Success*, 1951, p. 9.
4. Ibid., pp. 33-34.
5. United Nations Economic and Social Council, *Commission on Narcotic Drugs, Question of the Coca Leaf, Item 7 of the Previous Agenda of the 15th Session of the Commission, Lake Success*, 1959, p. 11.
6. *Time*, 12 April 1949, p. 44.
7. Ibid.
8. *Newsweek*, 12 June 1972, p. 124.
9. Charles Winick, "Use of Drugs by Jazz Musicians," in *Journal of Social Problems*, 1959, p. 242.
10. McLaughlin, "Cocaine: History or Regulation," p. 564.
11. Ibid.
12. Ibid.
13. *New York Times*, 1 February 1970, p. 18.

### Chapter 14. The 1960s and 1970s

1. Harry J. Anslinger, cited in *Bulletin on Narcotics*, July 1962, p. 530.
2. Jerry Hopkins, cited in Andrews and Solomon, *The Coca Leaf and Cocaine Papers*, p. 283.
3. Frederic Tuten, cited in *Easy Rider* screenplay, Peter Fonda, Dennis Hopper, Terry Southern, Raybert Productions, Inc., 1969, p. 37.
4. McLaughlin, "Cocaine: History and Regulation," p. 569.
5. Weil, "Gourmet Coca Taster's Tour of Peru," p. 47.
6. Richard Ashley, *Cocaine: Its History, Uses and Effects* (New York: St. Martin's Press, 1975), p. 161.
7 .Ibid.
8. Ibid.
9. Dr. Lester Grinspoon and James Balakar, *Cocaine and Its Social Evolution* (New York: Basic Books, 1976), p. 119.
10. Ibid., p. 179.

### Chapter 15. Cocaine Magic, Cocaine Myth

1. Zinburg and Robertson, *Drugs and the Public*, pp. 58–59.
2. Ibid., p. 58.
3. Nicholas Von Hoffman, "The Cocaine Culture: Current Rage for the Rich and Hip," in *Washington Post*, 23 April 1975, p. 75.
4. Jerry Hopkins, cited in Andrews and Solomon, *The Coca Leaf and Cocaine Papers*, p. 278.
5. *Newsweek*, 30 May 1977, p. 20.

### Chapter 16. Epilogue

1. Joel M. Hanna, "Use of Coca Leaf in Southern Peru: Adaptation or Addiction," *Bulletin on Narcotics*, vol. 29, No. 1, 1976.
2. Ibid., p. 71.
3. Nieschultz, cited in Andrews and Solomon, *The Coca Leaf and Cocaine Papers*, p. 274.

# Bibliography


Ackerknecht, Erwin H. *Medical Practices: Handbook of South American Indians.* Julian Stewart, editor. Washington, D.C.: Bureau of American Ethnology, 1949.

Acosta, José de. *The Natural History of the Indians.* London: Hakluyt Society, 1880.

Adams, Samuel Hopkins. "Great American Fraud." Colliers Magazine, 8 June 1907.

*American Medical Association, Journal of Nostrums and Quackery.* 2d ed. Chicago: JAMA, 1912.

Anderson, R. K. *Drug Smuggling and Taking in India and Burma.* Calcutta and Simla: Thacker and Spink Co., 1922.

Andrews, George, and David Solomon. *The Coca Leaf and Cocaine Papers.* New York and London: Harcourt Brace Jovanovich, 1975.

Anonymous. *Gourmet Cokebook.* White Mountain Press, 1972.

Arriaya, Father Pablo Joseph de. *The Extirpation of Idolatry in Peru.* Lexington: University of Kentucky Press, 1968.

Aschenbrandt, Dr. Theodor. *The Psychological Effect and Significance of Cocaine Muriate on the Human Organism.* In *Deutsche Medizinische Wochenshrift* 50, December 12, 1883; translated by Thérèse Byck, 1974.

Ashley, Richard. *Cocaine: Its History, Uses and Effects.* New York: St. Martin's Press, 1975.

Atkinson, D., M.D. *Magic, Myth and Medicine.* New York: World Publishing Company, 1956.

Baker, R., and M. Aldrich. *Historical Aspects of Cocaine Use and Abuse.* London: Fitz Hugh Ludlow Memorial Library, 1976.

Baudin, Louis. *A Socialistic Empire: Incas of Peru.* New York and London: D. Van Nostrand Co., Inc., 1961.

———. *Daily Life in Ancient Peru.* London: Allen and Unwin, Ltd. Museum St., 1961.

Baumann, Hans. *Gold and Gods in Peru.* London: Oxford University Press, 1963.

Becker, Hortense Koller. "*Carl Koller and Cocaine.*" *Psychoanalytic Quarterly* 32 (1963).

Benson, Elizabeth. *The Mochica.* New York and Washington: Frederick A. Praegar, 1972.

Bentley, Dr. W. H. "Erythroxoya Coca and the Opium and Alcohol Habits." *Detroit Therapeutic Gazette;* July 1878.

Bernfield, Sigmund. "Freud's Studies on Cocaine." *Journal of the American Psychoanalytic Association* 1, no. 4 (October 1953).

Blejer-Priesto, H., M.D. "Coca Leaf and Cocaine Addiction: Some Historical Notes." *Canadian Medical Journal* 93 (September 25, 1965).

Breecher, Edward M. *Licit and Illicit Drugs.* The Consumers' Union Report on Narcotics, Stimulants, Depressants, Inhalants, Hallucinogens and Marijuana—Including Caffeine, Nicotine and Alcohol. Consumers' Union of the United States, Inc. Boston and Toronto: Little, Brown and Co., 1972.

Brundage, Burr Cartwright. *Empire of the Incas.* Norman, Okla.: University of Oklahoma Press, 1963.

Burney, James. *A Chronological History in the South Seas and Pacific Ocean.* vol. 4, pp. 487–500. London: Kegan Paul, 1817.

Bushnell, G. H. S. *Peru.* New York and Washington: Frederick A. Praeger, 1957.

Byck, Dr. Robert. *The Cocaine Papers.* New York: Stonehill Press, 1974.

Chopra, I. C., and Chopra, Sir R. N. "The Cocaine Problems in India." *Bulletin on Narcotics,* April–June 1958.

Ciezea de Leon, Pedro. *The Incas of Pedro de Ciezea de Leon.* Von Hagen, ed. Norman, Okla.: University of Oklahoma Press, 1959.

Cooper, John M. "Stimulants and Narcotics." In *Handbook of South American Indians.* Julian Stewart, ed. Washington, D.C.: Bureau of American Ethnology, 1949.

Courtois-Suffit, Giroux R. *Traffic in Cocaine.* Paris: Académie Nationale de Médiécine, 1921.

Crowley, Aleister. *Diary of a Drug Fiend.* New York: Samuel Weiser, Inc., 1973.

Deneau, G. A. et al. "Self-Administration of Psychoactive Substances by the Monkey." *Psychopharmacologia* 16 (1969): 37.

Doyle, Arthur Conan. *The Sign of Four.* New York: Crown Publishers, 1967. (In *The Annotated Sherlock Holmes: The Four Novels and the Fifty-six Stories Complete.*)

Elsworth, Nancy, David Smith, et al., and Wesson, John, M.D. "Current Perspectives on Cocaine Use in America." *Journal of Psychedelic Drugs* 5, no. 2 (Winter 1972).

Esquemelling, John. *Buccaneers of America.* London: George Routledge and Sons, Ltd., 1684–85.

Federmann, Riennard. *The Royal Art of Alchemy.* Philadelphia: Chilton Book Company, 1964.

Fisher, Lillian E. *The Last Inca Revolt, 1780–1783.* Norman, Okla.: University of Oklahoma Press, 1966.

Flannery, Kent. "Origins of Agriculture." *Annual Review of Anthropology,* 1973.

Flornoy, Bertrand. *World of the Incas.* New York: Vanguard Press, 1956.

Freud, Sigmund. *On the General Effects of Cocaine.* (Lecture given at the Psychiatric Union on 5 March 1885.) Reprinted in *Cocaine Papers.* Edited by Robert Byck, M.D. New York: Stonehill Press, 1974.

———. *Über Coca.* Vienna: Cenralblatt für die ges. Therapil 2, 1885.

Fulop-Miller, René. *Triumph over Pain.* New York: The Bobbs-Merrill Co., 1938.

Gagliano, Joseph. "A Social History of Coca in Peru." Dissertation, Georgetown University, Washington, D.C., 1960.

Gaffney, George. "Narcotic Drugs—Their Origin and Routes of Traffic." *Drugs and Youth, Proceedings of the Rutgers Symposium on Drug Abuse,* 1969.

Garcilaso de la Vega. *El Inca: Royal Commentaries of the Incas.* Austin, Tex.: University of Texas Press, 1966.

Gay, G. R. et al. (C. W. Sheppard, D. S. Inaba, and J. A. Newmeyer). "Cocaine in Perspective: Gift from the Sun God to Rich Man's Drug." *Drug Forum* 2, no. 4 (Summer, 1973).

*Gentleman's Magazine* no. 84 (September 1814).

Goodman, Edward J. *Explorers of South America.* New York: Macmillan, 1972.

Goshen, Charles E. *Drinks, Drugs and Do-Gooders.* New York: Free Press, 1973.

Grinspoon, Dr. Lester, and Bakalar, James. *Cocaine and Its Social Evolution.* New York: Basic Books, Inc., 1976.

Grover, Norman. "Man and Plants against Pain." *Economic Botany* 19 (1965).

Guerra, Francisco, M.D. *The Pre-Columbian Mind.* London: Seminar Press, 1971.

Haile, H. G. *The History of Johann Faustus.* Urbana, Ill.: University of Illinois Press, 1965.

Hammond, William Alexander, M.D. in *Transactions of the Medical Society in Virginia,* From a volunteer paper, November 1887.

Hanna, Joel M. "Coca Leaf Use in Southern Peru: Some Biosocial Aspects." *American Anthropologist,* vol. 34, no. 2, (1974).

——— and Conrad A. Hornick. "Use of Coca Leaf in Southern Peru: Adaptation or Addiction." *Bulletin on Narcotics* 29, no. 1 (1976).

Hemmings, John. *Conquest of the Incas.* New York: Harcourt Brace Javanovich, 1970.

Hesse, Hermann. *Steppenwolf.* Translated by Basil Creighton. New York: Holt, Rinehart and Winston, 1964.

Hirschfield, Magnus. *Curious Sex Customs in the Far East.* New York: Grosset and Dunlap, 1935.

Holbrook, Stewart H. *Golden Age of Quackery.* New York: Macmillan, 1959.

Hopkins, Jerry. "Cocaine Consciousness: The Gourmet Trip." In George Andrews and David Solomon, *The Coca Leaf and Cocaine Papers,* New York: Harcourt Brace Jovanovich, 1975.

Humboldt, Baron von. *Personal Narrative of Travels of the Equinoctial Regions of the New Continent During the Years 1799–1804,* vol. 1. London: Kegan Paul, 1814.

Hymans, Edward and George Ordish. *The Last of the Incas.* New York: Simon and Schuster, 1963.

Jones, Ernest. *The Life and Work of Sigmund Freud.* New York: Basic Books, Inc., 1953.

Kahn, E. J., Jr. *The Big Drink: The Story of Coca-Cola.* New York: Random House, 1950.

Karsten, Rafael. *A Totalitarian State of the Past.* Port Washington, N.Y., and London: Kennikat Press, 1949.

King, Rufus. *The Drug Hangup—America's Fifty Years Folly.* New York: W. W. Norton and Co., 1972.

Koch, Christopher. "Cocaine." in *Literary Digest.* 28 March 1914.

———. "The Cocaine Habit." In *Literary Digest.* May 1911.

Kremers, Edward and R. Udang. *History of Pharmacy.* Philadelphia: Lippincott, 1963.

Lanning, Edward A. *Peru before the Incas.* Englewood Cliffs, N.J.: Prentice-Hall, 1970.

Liljestrand, G. "History of Local Anesthesia." In vol. 1 of *International Encyclopedia of Pharmacology and Therapeutics.* Elmsford, N.Y.: Pergamon Press, 1911.

Lumbreras, Luis G. *The People and Cultures of Ancient Peru.* Washington: Smithsonian Institution Press, 1974.

MacIntyre, Loren. *The Incredible Incas and Their Timeless Land.* Washington, D.C.: National Geographic Society, 1974.

MacNeish, Richard S. "Early Man in the Andes." In *Early Man in America.* Readings from *Scientific American,* San Francisco: W. H. Freeman and Company, 1971.

Mantegazza, Paolo. *Coca Experiences, Coca Leaf and*

*Cocaine Papers.* New York: Harcourt Brace Jovanovich, 1975.

Mariani, Angelo. *Coca and Its Therapeutic Action.* Translated by J. N. Jarnos. New York: privately printed by A. Mariani, 1890.

Marks, Geoffrey, and W. K. Beatty. *Medical Gardens.* New York: Charles Scribner's Sons, 1971.

Mason, Alden J. *The Ancient Civilization of Peru.* Baltimore, Md.: Penguin Books, 1957.

Martin, Richard T. "Role of Coca in the History, Religion and Medicine of South American Indians." *Economic Botany* 24 (1970): 422–37.

McLaughlin, Gerald T. "Cocaine: The History and Regulation of a Dangerous Drug." *Cornell Law Review* 58, no. 3 (March 1973).

Means, Phillip A. *Fall of the Inca Empire.* New York and London: Charles Scribner's Sons, 1932.

———. *The Spanish Main.* New York: Charles Scribner's Sons, 1935.

Mitchell, Charles L., M.D. Letter in "Nostrums and Quackery." *Journal of the American Medical Association,* vol. 2, 1912.

Moffet, Cleveland. "Rapid Increase of the most Dangerous Drug Habit; Menace of Cocaine-adulterated Soft Drinks and Catarrh Cures." *Hampton's Magazine* 24, no. 1 (1911).

Molina, Fernando Cabieses. "The Anti-Fatigue Action of Cocaine and Habituation to Coca in Peru. In George Andrews and David Solomon, *The Coca Leaf and Cocaine Papers.* New York: Harcourt Brace Jovanovich, 1975.

Moll, A A. *Aesculapius in Latin America.* Philadelphia: W. B. Saunders Co., 1944.

Moody, R. L. "Studies in Paleopathology Injuries to the Head among the Pre-Columbian Peruvians." *Annals of Medicinal History* 1 (1927).

Mortimer, William. *History of Coca.* Berkeley, Calif.:And/Or Press, 1914.

Musto, David F. *The American Disease—Origins of Narcotic Control.* New Haven, Conn.: Yale University Press, 1973.

Mutch, Dr. Nathan. "Cocaine." In *Guys Hospital Gazette* 46 (1932).

*National Clearinghouse for Drug Abuse Information,* Washington, D.C.: U.S. Government Printing Office, Series 11, no. 1, January 1972.

Nethercot, Anthur H. *Abraham Cowley: The Muses' Hannibal.* London: Oxford University Press, 1931.

*Newsweek.* "Smuggling: The Medicine Men." 12 June 1972, p. 37.

Osborne, Howard. *South American Mythology.* United Kingdom: Hamlyn Publishing Group, 1968.

Pitigrilli, Dino Segre. *Cocaine.* San Francisco: And/Or Press 1974.

Prescott, Frederich. *Control of Pain.* New York: Thomas Y. Crowell Co. 1964.

Proust, Marcel. *The Past Recaptured.* Translated by C. K. Scott Moncrieff. London: Chatto and Windus, 1952.

Rogers, David J. "Divine Leaves of the Incas." In *Natural History,* January 1963.

Rudorff, Raymond. *Belle Epoque.* London: Hamilton and Hamish, 1972.

Rusby, H. H. *Jungle Memories.* New York: McGraw-Hill Book Co., 1933.

Sauer, Carl O. "Cultivated Plants of South America." In *Handbook of South American Indians.* Julian Stewart, ed. Washington D.C.: Bureau of American Ethnology, 1949.

———. *Early Spanish Main.* Berkeley and Los Angeles, Calif.: University of California Press, 1966.

Scherzer, Karl D.C. *Voyage on the Novara. Narrative of the Circumnavigation of the Globe by the Austrian Frigate Novara.* London: Saunders, Otley and Co., 1861–63.

Schultz, Myron G., M.D., D.V.M. "The 'Strange Case' of Robert Louis Stevenson." JAMA 216, no. 1 (April 5, 1971). 216, no. 1.

Sharp, Gordon. "Coca and Cocaine Studied Historically." *Pharmaceutical Journal,* vol. 82, 1909.

Sitwell, Sacheverell. *Golden Wall and Mirador.* New York: The World Publishing Co., 1961.

Stephen-Hassard, Q. M. "Sacred Plant of the Incas." In *Pacific Discovery* 23, no. 5 (September–October 1970): 28.

Thomas, Norman. *Drug Plants That Changed the World.* London: George Allen, Ltd., 1958.

*Time Magazine.* "Peru: The White Goddess." April 1949, p. 44.

Towle, Margret A. *Ethnobotany of Pre-Columbian Peru.* New York: Viking Fund Publications in Anthropology, no. 30, 1961.

Towns, Charles B.D. "The Peril of the Drug Habit." In *Century Magazine,* 1912, p. 84.

Tuten, Frederic. Introduction to *Easy Rider.* Screenplay by Peter Fonda, Dennis Hopper, Terry Southern, Raybert Productions, Inc., 1969.

*United Nations Bulletin on Narcotics* 4, no. 4 (April–June 1952): 23.

United Nations Economic and Social Council (UNESCO). *Report on the 11th Session, Lake Success,* 1950.

United Nations Economic and Social Council (UNESCO). *Commission on Narcotic Drugs, 6th Session, Lake Success,* 1951.

United Nations Economic and Social Council (UNESCO). *Commission on Narcotic Drugs, Question of the Coca Leaf, Item #7 of the Previous Agenda of the 15th Session of the Commission, Lake Success,* 1959.

Von Hagen, Victor. *South America Called Them.* New York: A. A. Knopf, 1945.

Von Hoffman, Nicholas. "The Cocaine Culture: Current Rage for the Rich and Hip." *Washington Post,* 23 April 1975.

Waisbard, Simone and Roger. *Masks, Mummies and Magicians.* London: Oliver and Boyd, 1965.

Ward, G. A., Dr. "Cocaine" *Medical Record* 17 (1880): 497.

Watson, J. W., Col. "The Cocaine Problem." *New York Tribune.* 1 June 1903.

Weil, Andrew. "A Gourmet Coca Taster's Tour of Peru." *High Times,* May 1976. p. 37.

Williams, Edward H., M.D. "The Drug Menace in the South." In *Medical Record* 85 (1914).

Winick, Charles. "Use of Drugs by Jazz Musicians." In *Journal of Social Problems,* 1959.

Woods J.H. and Downs D.A. "The Psychopharmacology of Cocaine." In *Drug Use in America: Problem in perspective, the technical papers of the second report of the National Commission on marijuana and drug abuse.* Appendix, vol. 1. Washington, D.C.: U.S. Government Printing Office, 1973.

Yacovleff, E., and Herrera, F. L. "El Mundo Vegetal de Los Antigos Peruanos." *Revista del Museo Nacional* 3, no. 3 (1934).

Zimmerman, Arthur F. *The Reign of Viceroy Toledo, V^th^ Viceroy at Peru,* Caldwell, Idaho: Claxton Printers Ltd., 1938.

Zinberg, Norman E., and Robertson, John A. *Drugs and the Public.* New York: Simon and Schuster, 1972.

# Index